WITHDRAWN

# A Geometric Mechanism for Diffusion in Hamiltonian Systems Overcoming the Large Gap Problem: Heuristics and Rigorous Verification on a Model

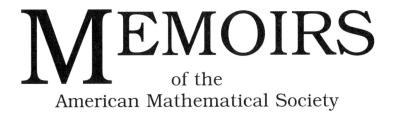

# MEMOIRS
of the
American Mathematical Society

Number 844

# A Geometric Mechanism for Diffusion in Hamiltonian Systems Overcoming the Large Gap Problem: Heuristics and Rigorous Verification on a Model

Amadeu Delshams
Rafael de la Llave
Tere M. Seara

January 2006 • Volume 179 • Number 844 (third of 5 numbers) • ISSN 0065-9266

**American Mathematical Society**
Providence, Rhode Island

2000 *Mathematics Subject Classification.*
Primary 37J40, 37C29, 34C37; Secondary 70H08, 37C50, 34C29.

**Library of Congress Cataloging-in-Publication Data**

Delshams, Amadeu.
   A geometric mechanism for diffusion in Hamiltonian systems overcoming the large gap problem : heuristics and rigorous verification on a model / Amadeu Delshams, Rafael de la Llave, Tere M. Seara.
      p. cm. — (Memoirs of the American Mathematical Society, ISSN 0065-9266 ; no. 844)
"Volume 179, number 844 (third of 5 numbers)."
Includes bibliographical references.
ISBN 0-8218-3824-5 (alk. paper)
   1. Nonholonomic dynamical systems.  2. Mechanics.  3. Differential equations—Qualitative theory.  I. Llave, Rafael de la.  II. Seara, Tere M., 1961– .  III. Title.  IV. Series.

QA3.A57  no. 844
[QA614.833]
510 s—dc22                                     2005053662
[515′.39]

# Memoirs of the American Mathematical Society

This journal is devoted entirely to research in pure and applied mathematics.

**Subscription information.** The 2006 subscription begins with volume 179 and consists of six mailings, each containing one or more numbers. Subscription prices for 2006 are US$624 list, US$499 institutional member. A late charge of 10% of the subscription price will be imposed on orders received from nonmembers after January 1 of the subscription year. Subscribers outside the United States and India must pay a postage surcharge of US$31; subscribers in India must pay a postage surcharge of US$43. Expedited delivery to destinations in North America US$35; elsewhere US$130. Each number may be ordered separately; *please specify number* when ordering an individual number. For prices and titles of recently released numbers, see the New Publications sections of the *Notices of the American Mathematical Society.*
    **Back number information.** For back issues see the *AMS Catalog of Publications.*
    Subscriptions and orders should be addressed to the American Mathematical Society, P. O. Box 845904, Boston, MA 02284-5904, USA. *All orders must be accompanied by payment.* Other correspondence should be addressed to 201 Charles Street, Providence, RI 02904-2294, USA.
    **Copying and reprinting.** Individual readers of this publication, and nonprofit libraries acting for them, are permitted to make fair use of the material, such as to copy a chapter for use in teaching or research. Permission is granted to quote brief passages from this publication in reviews, provided the customary acknowledgment of the source is given.
    Republication, systematic copying, or multiple reproduction of any material in this publication is permitted only under license from the American Mathematical Society. Requests for such permission should be addressed to the Acquisitions Department, American Mathematical Society, 201 Charles Street, Providence, Rhode Island 02904-2294, USA. Requests can also be made by e-mail to reprint-permission@ams.org.

*Memoirs of the American Mathematical Society* is published bimonthly (each volume consisting usually of more than one number) by the American Mathematical Society at 201 Charles Street, Providence, RI 02904-2294, USA. Periodicals postage paid at Providence, RI. Postmaster: Send address changes to Memoirs, American Mathematical Society, 201 Charles Street, Providence, RI 02904-2294, USA.

# Contents

# Abstract

We introduce a geometric mechanism for diffusion in a priori unstable nearly integrable dynamical systems. It is based on the observation that resonances, besides destroying the primary KAM tori, create secondary tori and tori of lower dimension. We argue that these objects created by resonances can be incorporated in transition chains taking the place of the destroyed primary KAM tori.

We establish rigorously the existence of this mechanism in a simple model that has been studied before. The main technique is to develop a toolkit to study, in a unified way, tori of different topologies and their invariant manifolds, their intersections as well as shadowing properties of these bi-asymptotic orbits. This toolkit is based on extending and unifying standard techniques. A new tool used here is the scattering map of normally hyperbolic invariant manifolds.

The model considered is a one-parameter family, which for $\varepsilon = 0$ is an integrable system. We give a small number of explicit conditions the jet of order 3 of the family that, if verified imply diffusion. The conditions are just that some explicitely constructed functionals do not vanish identically or have non-degenerate critical points, etc.

An attractive feature of the mechanism is that the transition chains are shorter in the places where the heuristic intuition and numerical experimentation suggests that the diffusion is strongest.

Received by the editor May 9, 2003.

2000 *Mathematics Subject Classification.* Primary 37J40; 37C29 , 34C37; Secondary, 70H08, 37C50, 34C29,

*Key words and phrases.* Arnol'd diffusion, instability, heteroclinic connections.

# CHAPTER 1

# Introduction

The phenomenon of diffusion in phase space for Hamiltonian systems is important for applications and has attracted a great deal of attention both in the mathematical and in the physical literature. The rather loose (and somewhat inadequate) name *diffusion* captures the intuition that there are trajectories that wander widely and explore large regions of phase space. Even if diffusion is, presumably, a phenomenon that happens in rather general systems—as conjectured in [**Arn63b**], particular attention has been given to studying it in mechanical systems close to integrable. On the one hand, one hopes that the good understanding we have of integrable systems can be transferred to quasi-integrable systems. On the other hand, since integrable systems do not exhibit any diffusion, quasi-integrable systems—which also present KAM tori and other obstacles to diffusion—are a good test case for the generality of the phenomenon.

In the mathematical literature, there are different precise definitions of diffusion trying to capture the idea of "large" excursions. However, the dominant geometric paradigm for diffusion near integrable systems has been till recently the mechanism proposed by V. I. Arnol'd and documented, for example, in [**Arn64, AA67**]. This mechanism, henceforth referred as *classical Arnol'd diffusion*, is based on the existence of chains of whiskered tori—remnants of those of the integrable system—and such that the unstable manifold of one intersects the stable manifold of the next one. The classical Arnol'd diffusion has not been shown to be a generic phenomenon. On the other hand, several of its ingredients have been verified for some systems [**CG94, CG98, Moe96**] and some special families exhibiting it have been constructed [**Dou88, DLC83, Gal99, FM01**].

In this paper, however, we want to argue that the mechanism described in [**Arn64**] is not the only mechanism for diffusion in quasi-integrable systems and that there are other geometric mechanisms that can be studied with mathematical rigor. Of course, there are other possible approaches besides geometric perturbation theory, notably, variational methods which have produced significant progress in recent years, but we will not deal with them.

The first goal of this paper is to formulate heuristically two variants of the classical Arnol'd mechanism in which the diffusion takes place through orbits that follow transition chains of whiskered tori as in [**Arn64**]. Nevertheless, the transition chains we consider contain objects which are not the KAM tori present in the original system.

In the first of the mechanisms we propose, the whiskered tori involved in the transition chain are not only those tori that can be studied as perturbations of tori in the original system (called primary tori). In our mechanism, we incorporate tori (called secondary tori) which cannot be continuously deformed into tori invariant under the original system. They are topologically different from the tori present in the unperturbed system (see Definition 2.2). The generation of secondary tori due to resonances has been discussed in [**LW04**].

In the second mechanism we present here, some of the whiskered tori which enter in the transition chain are of lower dimensions, e.g, periodic orbits. Indeed, they are tori that survive resonances in which the primary tori are destroyed by a mechanism similar to that considered in [**Poi99**, Chap. V, §81], [**Tre91**].

As we will see, the new objects (secondary tori and lower dimensional tori with their manifolds) dovetail very precisely in the gaps between KAM tori (see Chapter 8 and Figure 8.1 on page 39). Hence, they can be used as elements of transition chains that overcome the large gap problem.

The second goal, which takes up the bulk of this paper, is to verify rigorously the existence of these mechanisms in rather concrete systems. The verification will be rather explicit, and given a concrete systems there are finite calculations which establish it. In particular, we will study a model that has been studied already in [**HM82**]. We note that the model we discuss presents the *large gap problem* (namely that the size of the gaps between the KAM tori is larger than the size of first order change in the (un)stable manifolds) that has played an important role in Arnol'd diffusion. An excellent discussion of the large gap problem and, indeed of the problem of diffusion can be found in [**Moe96**].

Our main rigorous result Theorem 4.1 establishes that the mechanisms that we present overcome the large gap problem in the model considered. Theorem 4.1 presents several rather explicit sufficient conditions that ensure that a system which verifies them has orbits that traverse a large gap region.

We believe that the mechanisms proposed here have the advantage that they fit better some of the intuition gathered from numerical and real experiments than the mechanism of [**Arn64**]. Note that the numerical, experimental and geometric intuition is that resonances generate diffusion. (See e.g. the classical [**Chi79**] or [**CSUZ89, Mei92, JVMU99, Las93, LR02**], among many others.) On the other hand, the mechanism of [**Arn64**] has difficulty dealing with resonances, which destroy the primary tori. Indeed, one of the main problems to establish rigorously the existence of the mechanism of [**Arn64**] is the *large gap problem*, which refers to the fact that the resonances create gaps in the families of primary whiskered tori whose size is bigger than the angle between the whiskers of the primary whiskered tori. In other words, the mechanism of [**Arn64**] for diffusion is difficult to verify—and can be presumably false—precisely in the places where experimental evidence suggests that diffusion should be most intense.

For the mechanisms that we propose here, we observe that resonances, even if they destroy the primary tori, they create secondary tori and tori of lower dimension that bridge them, so that the transition chains can continue. Indeed, in agreement with the physical intuition, the secondary tori created by resonances lead to a larger increase in action in their elements of the transition chain. (See [**Hal97, Hal99**] for a discussion of the role of double resonances in diffusion.)

We are far from believing that the mechanisms we present here are the only mechanisms for diffusion. Some other geometric mechanisms have been rigorously established for other systems. For example, [**DLS00, BT99, DLS05**] study geometrically a system that has been studied in [**Mat95**] by using variational methods. A variational approach to Arnol'd diffusion can be found in [**BB02, BBB02, BBB03**] and announcements of other variational methods are in [**Xia98, Mat02**]. The papers [**Lla02, Moe02, EMR01, Tre02a, Tre02b, Tre04**] study other geometric mechanisms. There are heuristic descriptions and numerical explorations of other geometric mechanisms in [**LT83, Ten82, CLSV85**].

In Chapter 2, we will describe the proposed mechanism in an informal and conjectural way. Hence, the reader interested only in results that are rigorously proved can skip this chapter altogether. Then, in Chapter 3, we present a class of model systems in which we will verify rigorously the mechanisms described in Chapter 2. In Chapter 4, we will state Theorem 4.1 that verifies the mechanism in the concrete model introduced in Chapter 3.

The subsequent chapters are devoted to the proof of Theorem 4.1. In Chapter 5 we introduce some notation. In Chapter 6, we describe the geometry of the problem, which has several elements in common with that of [**DLS00**]. Even if the present paper is logically independent of [**DLS00**] (we only use a few of the technical results), we have included in Section 12.2 a discussion of the similarities and differences between the phenomena considered and the methods used in this paper and in [**DLS00**]. In subsequent sections we will prove Theorem 4.1, following the steps indicated in Chapter 2, and we will develop tools to establish that, indeed, a given system satisfies the proposed mechanism.

We note that the methods we present are rather constructive and are capable of establishing—or at least reduce to a finite calculation—the existence of the mechanisms in a concrete system. The verification of the mechanism amounts just to computation of explicitly given formulas and verifying that they do not vanish. Indeed, in Chapter 13, we present some systems where it is possible to compute the formulas needed in closed form so that we can ensure that these systems have trajectories that cross the resonance gaps.

We call attention to the fact that the proof we present is quite modular and consists on well defined steps that depend only on the results of the previous ones. If one is willing to make assumptions that yield the results of one step, or finds a technique that verifies the steps independently, then the rest of the proof can be used.

We also note that most of the tools we need for the verification of the mechanism proposed here are variants of standard techniques in dynamical systems and specially in diffusion. Most of the results we use can be obtained through readily available techniques. Hence, the main novelty of this paper lies in the overall strategy and the dovetailing of different geometric structures. This is why we have included a heuristic description of the mechanism. Undoubtedly, given the heuristic description, some experts will have little trouble filling the details themselves.

We hope that these improvements, sharpening the analytic tools so that they become reusable as parts of longer arguments, could be useful for the study of other mechanisms for diffusion. We also expect that the ingredients we have studied—mainly normally hyperbolic invariant manifolds, secondary tori—can be rearranged in other ways to provide additional mechanisms for diffusion or that they can be integrated with other approaches, notably variational methods. But we note that convexity of the problem does not play any role in our methods in contrast with variational method.

One fundamental tool that we adopted from [**DLS00**] is the use of scattering maps on normally hyperbolic invariant manifolds. The scattering map method allows us to discuss in a geometric way heteroclinic phenomena. In particular, we can discuss comfortably the intersection of invariant manifolds of tori of different topologies and even of different dimensions. This discussion does not seem straightforward in the usual Melnikov theory which often requires that one puts both objects experiencing the homoclinic phenomena in the same system of coordinates.

We have chosen to verify the results for one-parameter families and formulate our results in such a way that these results apply for all $|\varepsilon| < \varepsilon^*$. Of course, the $\varepsilon^*$ depends on the family and can tend to zero as the family considered approaches a particularly degenerate case which does not satisfy the assumptions of the theorem. We have also chosen to formulate our results by obtaining computable sufficient conditions on the family for the results to apply.

For one-parameter families, it is customary to classify them—following the notation of [**CG94**]— in *a-priori stable* or *a-priori unstable* systems depending on whether the unperturbed system is strongly integrable (can be put in action-angle variables) or not. Of course, this is not the only possible point of view and it is interesting to compare this formulation with other formulations of the problem of diffusion. For example, the method in [**Arn64**] considers families depending on two parameters (the second one exponentially small with respect to the first).

It is also natural and customary to make assertions for generic or for "cusp residual sets" (see [**Mat02**]). Nevertheless, in this paper we have adopted the concrete verification point of view and relegated a tentative discussion of genericity to remarks along the text and to Section 12.3. Since all the conditions that we need are basically that rather explicit expressions do not vanish, it is clear that the results apply to systems that are generic in many precise senses.

It is important to realize that, when one considers 2-parameter families—a fortiori if one consider generic results—the distinction between a-priori stable systems and a-priori unstable systems does not make sense. One can use the first parameter to move the system away from a-priori stable and then use the second parameter—or genericity—applying the methods of a-priori unstable systems. Hence, one can hope that the methods of one-parameter families near a-priori unstable systems apply to obtain two parameters or generic results.

The study of one-parameter families near a-priori stable systems seems to require considerations of another nature. There are indeed normal forms near resonances similar to the systems we consider here. Nevertheless there appear exponentially small phenomena in the parameter and dealing with them without introducing other exponentially small parameters or generic perturbations requires delicate analytical tools.

In the long time that this paper was under editorial consideration, an announcement of the main results of this paper appeared in [**DLS03**]. We hope that this announcement can serve as a reading guide for this paper. During this time, we also become aware of the following papers on related topics [**Ber04, BK04, CY04b, CY04a, DLS04, GL05, Kal03, KMV04, MS04**]. Some of these papers provide alternative methods for some of the steps of this paper and some give at least partial answers to questions raised at the end such as times of diffusion. We hope in the future to make a more detailed discussion.

# CHAPTER 2

# Heuristic discussion of the mechanism

In this chapter, we will describe heuristically what are the ideas that lead to the mechanism of diffusion proposed in this paper.

The discussion in this chapter will be completely non-rigorous. Nevertheless, we hope that it can serve as stimulus for future work. Of course, the reader interested only in rigorous results can skip to Chapter 3, and the following after browsing through the definitions introduced in Section 2.1.

## 2.1. Integrable systems, resonances, secondary tori

We start by collecting some standard definitions and setting the notation we will use.

For our purposes, we describe a *strongly integrable system* as a system which has phase space $\mathbb{R}^d \times \mathbb{T}^d$, where $\mathbb{T} = \mathbb{R}/2\pi\mathbb{Z}$, and on which the motion is given by:

$$(2.1) \qquad\qquad \Phi_t(I, \varphi) = (I, \varphi + \omega(I)t)$$

in the case of flows or by:

$$(2.2) \qquad\qquad F(I, \varphi) = (I, \varphi + \omega(I))$$

in the case of maps.

According to this rather restrictive definition, the mathematical pendulum, a system in $\mathbb{R}^1 \times \mathbb{T}^1$ described by the Hamiltonian

$$P(p, q) = p^2/2 + \cos q - 1,$$

is not a strongly integrable system even if it has a conserved quantity—the Hamiltonian $P$—and all the motions are quite orderly. Some of the motions consist on librations, some others are rotations and the rotations and librations are separated by orbits that start and end in the critical point $p = 0, q = 0$. Note that these orbits (usually called *separatrices*) separate two different topological types of orbits. Hence there is no hope of writing global action-angle coordinates that straddle them.

Consequently one may introduce the less stringent definition of *integrable system* in which the representations (2.1) or (2.2) hold in open dense sets. These open sets are delimited by special submanifolds called *separatrices*. These separatrices are, at the same time, stable and unstable manifolds of lower dimensional tori invariant for the flow. Following [**CG94**], it is customary to refer to systems as (2.1), (2.2) as *a priori stable*. Otherwise, systems which present separatrices and hyperbolicity, but which are strongly integrable piecewise in the complement of the separatrices, are called *a priori unstable*.

REMARK 2.1. There are systems that present separatrices which are not the stable or unstable manifolds of other hyperbolic sets (e.g., take a pendulum with potential $V(q) = (\cos q - 1)^4$ so that the critical points are not hyperbolic points). Such situations seem not to have received much attention in the mathematical literature even if they appear naturally in applied models. Nevertheless, see [**BF04**].

A *quasi-integrable system* is a system which is close to an integrable system in the topology of a space of sufficiently smooth functions. Sometimes, it will be convenient to consider families indexed by a parameter $\varepsilon$ so that the system for $\varepsilon = 0$ is integrable. For $|\varepsilon|$ small enough, the system will be quasi-integrable.

We recall that, under appropriate differentiability and non-degeneracy conditions, if the unperturbed system is strongly integrable, the KAM theorem shows that a large fraction of the measure of space of a quasi-integrable system is covered by invariant tori with a Diophantine frequency $\omega(I)$.

The KAM theorem does not deal with regions on which

$$(2.3) \qquad \omega(I) \cdot k = \mathrm{O}(\varepsilon^{1/2}), \quad k \in \mathbb{Z}^d \setminus \{0\}.$$

The power $1/2$ is optimal as can be seen in examples such as $\frac{I^2}{2} + \varepsilon \sin \varphi$, and it was established in [**Sva80, Neĭ81, Pös82**]. The $\mathrm{O}(\varepsilon^{1/2})$ in the size that needs to be excluded is non-uniform in $k$ and goes to zero when $|k| \to \infty$.

We will call *resonances* the values of $I$ for which $\omega(I) \cdot k = 0$, for some $k \in \mathbb{Z}^d \setminus \{0\}$. We will call *resonant regions* the regions (2.3) around the resonance that need to be excluded in the proof of KAM theorem.

It is known empirically that very often, in these resonant regions there are dynamical objects which are not present in the integrable system. Typically, the resonant regions contain poorly understood areas called *the chaotic sea* which includes homoclinic intersections of lower dimensional orbits and secondary tori. Indeed, secondary tori are the most visible objects in numerical simulations. In two dimensions, when the visualization is very clear, these secondary tori are the tori in the *islands* in the chaotic sea. (See Figure 8.1.)

More precisely:

DEFINITION 2.2. *We say that a $(d - k)$-dimensional torus invariant under the flow is a secondary torus when it is retractable to a set $\{I\} \times \{(\varphi_1, \ldots, \varphi_l)\} \times \mathbb{T}^{d-l-k}$.*

The fact that resonances generate secondary tori can be established rigorously under suitable non-degeneracy assumptions on the integrable system and the perturbation. We will develop one such proof in Section 8.5.

We also point out that besides the $d$-dimensional tori, KAM theory can establish the persistence of $d - k$-dimensional tori, $k < d$, when they have $k$

stable and unstable directions. These tori will be called whiskered tori (see [**Gra74, Zeh76**]). Optimal measure estimates for these tori are available from [**Val00**]. In this work, besides the primary whiskered tori considered in KAM theory, we will consider also secondary whiskered tori, that is, whiskered tori that cannot be deformed into tori present in the unperturbed system.

It is also known that resonances generate lower dimensional whiskered tori. (See, for instance, [**Poi99**, Cap V, §81], [**Tre91, Nie00, DG01**] for primary tori and [**LW04**] for secondary tori.)

In our study of diffusion, there will not be much difference between whiskered tori and fully dimensional tori. We will use the theory of normally hyperbolic invariant manifolds to find invariant submanifolds. The maximal KAM tori in these invariant submanifolds will correspond to whiskered tori for the full system. This observation was already done in [**Moe96, DLS00**] and it is exploited in [**Sor02**],which gives a proof of persistence of whiskered tori by combining normally hyperbolic manifolds and the proof of the persistence of maximal tori. On the other hand, distinction between primary and secondary tori will be very important for us.

## 2.2. Heuristic description of the mechanism

The systems we consider will be perturbations of integrable ones. To fix ideas, we will consider an unperturbed system which is integrable but not strongly integrable (i.e. it is a priori unstable), and we will perturb it.

Our unperturbed system will admit action-angle variables in open sets divided by *separatrices* i.e. invariant manifolds that end in lower dimensional tori (of codimension one). We will refer to these tori present in the unperturbed system as *primary whiskered tori*.

An important example of these systems, which has been considered many times in the literature and which we will revisit here, is the system consisting of a pendulum and an oscillator uncoupled. Note that this system exhibits a behavior which is very similar to the behavior we will find in a neighborhood of a resonance in a typical system.

When we consider all the whiskered tori of the integrable system, we obtain a normally hyperbolic invariant manifold (which we call the resonant manifold). In the integrable case, we note that the stable and unstable manifolds of the resonant manifold agree. Note also that if we consider the motion of the integrable system restricted to the resonant manifold, it is strongly integrable, that is, it is described by global action-angle variables. The whiskered tori of the full system are KAM tori inside the resonant manifold. The motion restricted to the resonant manifold presents resonances, which in turn correspond to the double resonances of the original system.

As it is well known to specialists in Arnol'd diffusion, adding a perturbation of size $\varepsilon$ preserves the resonant manifold and its stable and unstable invariant manifolds. For typical perturbations the foliation by tori in the perturbed manifold persists except for gaps or order $O(\varepsilon^{1/2})$. Moreover, the stable and unstable manifolds move by an amount of order $O(\varepsilon)$. It is rather straightforward to show that when there are tori at a distance $O(\varepsilon)$, under appropriate non-degeneracy conditions on the perturbation, one can easily show that there exist heteroclinic intersections between the tori and, hence, construct transition chains along the resonant manifold. However, near the gaps of order $O(\varepsilon^{1/2})$, these leading order considerations do not allow to conclude existence of transition chains. This is what is called the *large gap problem*.

The main idea of the method proposed here is to study carefully the objects generated by the double resonance. As it is more or less folklore, the resonances destroy the primary whiskered tori present in the original system but create objects of other types. In particular, they create secondary whiskered tori and whiskered tori of lower dimension.

The key point of the proposed mechanism is that the secondary whiskered tori and the stable and unstable manifolds of the lower dimensional tori come very close to the region covered by the primary tori. The secondary tori *bridge* the resonant region. Hence, by incorporating them in the transition chains we can overcome the large gap problem.

The heuristic description of the mechanism proposed here to overcome the large gap problem is:

1) Outside the regions of double resonance, KAM theorem applies, and we obtain the existence of codimension one whiskered tori which are extremely close to each other. Hence one can make transition chains following the classical Arnol'd mechanism, that is, chains formed by heteroclinic connections between the stable and unstable manifolds of primary whiskered tori.

2) Inside the regions of double resonance, and very close to their boundary, we can find KAM secondary whiskered tori.

2') Inside the regions of double resonance, and very close to their boundary, we can find also stable and unstable manifolds of whiskered tori of lower dimension than those of the integrable one (indeed, periodic orbits in our model).

3) The secondary whiskered tori, the lower dimensional whiskered tori and the primary whiskered tori lie on a normally hyperbolic invariant manifold which is a continuation of the resonant manifold for the integrable system.

4) This normally hyperbolic invariant manifold has stable and unstable invariant manifolds which, under some explicit non-degeneracy conditions on the perturbation, intersect transversally.

5) Under some explicit non-degeneracy conditions on the perturbation it is possible to produce transition chains, that is sequences of invariant whiskered tori in which the unstable manifold of one intersects transversely the stable manifold of the next. These transition

chains may involve primary whiskered tori, secondary whiskered tori and lower dimensional whiskered tori.

6) Once we have the transition chains—which may involve secondary tori or lower dimensional tori—it is possible to use variants of the usual *"obstruction property"* mechanism to show that there are orbits that follow the transition chains.

Perhaps the main conceptual novelty lies in point 2) and 2') above where we identify geometric structures in the double resonances which allow to bridge them.

Note that our mechanisms depend on establishing the existence of transverse intersections of stable and unstable manifolds for objects of different topological type. In such a case, one cannot easily use the methods of standard Melnikov theory which rely on using a similar coordinate representation for both of them. A technical tool which could be of independent interest is the scattering map, defined on the normally hyperbolic invariant manifold thanks to the fact that its stable and unstable invariant manifolds intersect transversally, and which allows us to discuss the heteroclinic intersection of invariant objects of different type.

It will follow from the explicit formulas used to verify the hypotheses in the steps above (specially 4), 5) ) that the conditions are satisfied in generic set of perturbations.

REMARK 2.3. We will see that the resonant regions, even if they have a size $O(\varepsilon^{1/2})$, are connected by only one element in the transition chain, whereas outside of the resonant regions, one step of the transition chain only makes a much smaller step in space.

If one makes the conjecture (see [**Chi79**], amplified in [**CV89**]) that the time spent in a transition chain is roughly proportional to the number of tori in the transition chain—there could be a factor $|\log \varepsilon|$—one obtains that the resonant region will be crossed over faster than other regions in the transition chain. Hence, the diffusion will be a slow drift punctuated by much larger jumps when the system approaches a resonance. This fits well with the descriptions in the empirical literature, in particular, the literature on Levy flights [**SZF95, Zas02**] or the experimental observations that confirm that diffusion is much more apparent near resonances [**Las93, LR02**]. Of course, the current state of the art is far from a proof of this conjecture. In particular, there do not seem to exist techniques to produce a rigorous statistical theory. Nevertheless we find encouraging that our methods of proof move in the direction suggested by intuition and experimental observations, and show some semblance of agreement with them.

REMARK 2.4. In many physical applications, one has to consider systems which consist of many identical sub-systems interacting by a local coupling. In [**HL00**] there is an empirical and heuristic study of the abundance of secondary tori in these coupled oscillators. The conclusion of numerical experiments and heuristic arguments of [**HL00**] is that secondary tori are much more abundant than KAM primary tori in systems of coupled oscillators.

From the mathematical point of view, in this paper we will content ourselves with studying just one class of models. The models have been chosen because the verification of the mechanism is non-trivial, but devoid of technical complications. The class of models we consider were introduced in the paper [**HM82**], which ignored the large gap problem.

Most of the methods that we employ in our verification have been standard tools of the geometric approach to standard Arnol'd diffusion. Nevertheless, we call attention to the fact that we also make good use of the perturbation theory of normally hyperbolic invariant manifolds to provide an skeleton along which diffusion takes place. This is a common feature with [**DLS00, DLS05**], which consider yet another mechanism of diffusion and with [**Moe96**].

The main object that organizes the mechanism here is a normally hyperbolic invariant manifold $\tilde{\Lambda}_\varepsilon$ close to the resonances. The stable and unstable manifolds of this manifold will intersect transversally. This will allow us to define two dynamics on the manifold $\tilde{\Lambda}_\varepsilon$. One is the dynamics restricted to the manifold, which we called the *inner map* in [**DLS00**]. Following again [**DLS00**], we will also define a *scattering map* in Chapter 9. Given an orbit $\tilde{z}(t)$ that performs a homoclinic excursion to $\tilde{\Lambda}_\varepsilon$, we can find two orbits in $\tilde{\Lambda}_\varepsilon$ that approach $\tilde{z}(t)$ in the future and in the past. The scattering map associates the orbit asymptotic in the past to the orbit asymptotic in the future. It will be shown that, when one considers all the orbits $\tilde{z}(t)$ chosen in a homoclinic connection, we obtain a scattering map which is smooth.

One can obtain diffusing pseudo-orbits by applying alternatively the inner map and the scattering map. The iterations of the inner map stay in invariant objects—primary and secondary KAM tori or stable manifolds of lower dimensional tori. This corresponds to orbits that perform a homoclinic excursion when the perturbation in favorable but that, otherwise stay "parked" near $\tilde{\Lambda}_\varepsilon$. The reason why we can construct these orbits is because the secondary objects dovetail inside the gaps created in the foliation of primary tori. The bulk of Chapter 8 is devoted to showing the existence of these objects and obtaining detailed estimates on their geometry.

Since the motion generated by the inner map contains invariant objects such as tori, it is possible to use obstruction properties (see Chapter 11) to construct orbits that diffuse. There is a large literature on obstruction properties. Not all the methods proposed in the literature adapt to our situation where one needs to consider tori of different topologies. Nevertheless, we will verify that some of the methods do apply. We will follow the method explained in [**DLS00**]. The main ingredient of this method is the Lambda Lemma result in [**FM00**], which does not depend neither on the topology of the embedding of the torus nor on the dimension of the torus.

It is very likely that there are other methods to verify the mechanism presented here and indeed other mechanisms for diffusion. We expect that many of the results established here could be established for concrete systems using variational or more topological methods or proved for generic systems by other methods.

# CHAPTER 3

# A simple model

In the rest of the paper we will verify rigorously the mechanism presented above in a specific model. The model has been chosen in such a way that the verification of the proposed mechanism is as simple as possible among the non-trivial ones.

This model has been studied in several other places, but to the best of our knowledge, the mechanism described in this paper has not been considered. In particular, it was considered in [**HM82**], but the large gap problem was not addressed in that paper. We also note that the model we consider is a standard model of the behavior near resonances.

Even if the analysis in this paper will require several unduly restrictive assumptions, we hope that it could lay the foundations for further progress.

We will consider a mechanical system described by the non-autonomous Hamiltonian, periodic in time

$$\begin{aligned} H_\varepsilon(p,q,I,\varphi,t) &= H_0(p,q,I) + \varepsilon h(p,q,I,\varphi,t;\varepsilon) \\ &= P_\pm(p,q) + \frac{1}{2}I^2 + \varepsilon h(p,q,I,\varphi,t;\varepsilon) \end{aligned}$$

(3.1)

where we denote by

(3.2)
$$P_\pm(p,q) = \pm\left(\frac{1}{2}p^2 + V(q)\right)$$

and where $V(q)$ is a $2\pi$-periodic function. We will refer to $P_\pm(p,q)$ as the *pendulum*. (It is a physical pendulum when we take $P_+$, and $V(q) = \cos q - 1$, which is a good example to keep in mind.) The term $\frac{1}{2}I^2$ of (3.1) describes a rotator whose frequency changes with the energy of the oscillation (equivalently with the action).

The final term $h$ in (3.1) describes a small coupling between the rotator and the pendulum that depends periodically on time. We will assume—besides differentiability properties—that $V$ reaches a maximum at one point (we will assume without loss of generality that this point is 0) and that the maximum is non-degenerate (i.e. $V''(0) < 0$). We will assume for the sake of simplicity of exposition that 0 is the only local maximum. This later assumption can be eliminated just by complicating the notation.

REMARK 3.1. The choice of sign in $P_\pm$ will not make any difference in our arguments which are based on hyperbolicity and KAM theory. Note however that the Hamiltonian with $P_+$ is convex for large $I, p$ but with $P_-$ is neither convex nor positive definite.

The assumption of positive definiteness seems to be very important for variational methods [**Xia98, Mat02**]. On the other hand, we note that the physical intuition [**Chi79, TLL80**] is that systems with subsystems of different signatures tend to be more unstable.

It will be convenient to consider the system (3.1) as described by an autonomous Hamiltonian flow on $(\mathbb{R} \times \mathbb{T})^3$—which we will call the symplectic extended phase space—endowed with the standard symplectic structure. The autonomous Hamiltonian will be:

$$
\begin{aligned}
(3.3) \qquad \tilde{H}_\varepsilon(p, q, I, \varphi, A, s) &= A + H_\varepsilon(p, q, I, \varphi, s) \\
&= A + H_0(p, q, I) + \varepsilon h(p, q, I, \varphi, s; \varepsilon) \\
&= A + P_\pm(p, q) + \frac{1}{2}I^2 + \varepsilon h(p, q, I, \varphi, s; \varepsilon).
\end{aligned}
$$

We will follow the standard convention of assuming that the pairs $(p, q) \in \mathbb{R} \times \mathbb{T}$, $(I, \varphi) \in \mathbb{R} \times \mathbb{T}$, and $(A, s) \in \mathbb{R} \times \mathbb{T}$, are symplectically conjugate variables.

Note that $h$ does not involve $A$ so that the equations of motion of the pair $(A, s)$ are just

$$
\dot{A} = -\partial_s \tilde{H}_\varepsilon(p, q, I, \varphi, s) = -\varepsilon \frac{\partial h}{\partial s}(p, q, I, \varphi, s; \varepsilon), \qquad \dot{s} = 1.
$$

The introduction of the extra variables $(A, s)$ is a standard device to formulate periodic in time perturbations as an autonomous system. The extra variable $s$ makes the system autonomous and the variable $A$ is symplectically conjugate to $s$ to be able to treat the resulting system as a Hamiltonian one. So, even if the system described by (3.3) is, strictly speaking, a three degrees of freedom system, we refer to it as a two and a half degrees of freedom system.

Moreover, the variable $A$ does not play any dynamical role. Note that $A$ does not appear in any of the differential equations for any of the coordinates, including itself. Then, one can study the dynamics of the variables $(p, q, I, \varphi, s)$, and then $\dot{A} = -\partial_s \tilde{H}_\varepsilon(p, q, I, \varphi, s)$ is just a quadrature. Hence, we will consider the equations:

$$
\begin{aligned}
\dot{p} &= \mp V'(q) & -\varepsilon\frac{\partial h}{\partial q}(p,q,I,\varphi,s;\varepsilon) \\[2mm]
\dot{q} &= \pm p & +\varepsilon\frac{\partial h}{\partial p}(p,q,I,\varphi,s;\varepsilon) \\[2mm]
\dot{I} &= & -\varepsilon\frac{\partial h}{\partial\varphi}(p,q,I,\varphi,s;\varepsilon) \\[2mm]
\dot{\varphi} &= I & +\varepsilon\frac{\partial h}{\partial I}(p,q,I,\varphi,s;\varepsilon) \\[2mm]
\dot{s} &= 1.
\end{aligned}
$$

(3.4)

We will denote by $\mathcal{H}_\varepsilon$ the vector field (3.4) generated by the Hamiltonian of (3.3). We will denote by $\Phi_{t,\varepsilon}(\tilde{x})$ the flow generated by the vector field $\mathcal{H}_\varepsilon$ for $\tilde{x}$ in the extended phase space $(\mathbb{R}\times\mathbb{T})^2\times\mathbb{T}$.

REMARK 3.2. Notice that introducing the extra variable $A$ makes some difference in the geometric nature of the objects. For example, if we find a KAM torus in the non-autonomous system, it will become a family of tori—indexed by the variable $A$—in the autonomous system. Hence, even if the KAM tori in the non-autonomous system are, for typical perturbations, a Cantor set of tori, in the autonomous version they always include one-parameter families indexed by the variable $A$.

The Hamiltonian $H_0(p,q,I)$ in (3.1) will be referred to as the unperturbed Hamiltonian. It is customary to describe $H_0$ as integrable. As we pointed out in Section 2.1, the Hamiltonian system of Hamiltonian $H_0$ is a priori unstable. Indeed, the system described by $P_\pm$ presents different topological types of oscillations which are separated by special orbits called separatrices ending with zero velocity on the maximum of $V$.

# CHAPTER 4

# Statement of rigorous results

In this chapter we formulate the main Theorem of this paper. Actually, we will prove somewhat more general results which we formulate later when we have introduced more notation. For example besides orbits which transverse the gaps, our proof also establishes the existence of a symbolic dynamics. We will also indicate results which use slightly weaker hypotheses taking advantage of the existence of several homoclinic intersections.

We will consider a neighborhood $\mathcal{S} \subset \mathbb{R} \times \mathbb{T}$ of the separatrix of the pendulum we are studying, and consider the compact set

(4.1) $$\mathcal{D} := \mathcal{S} \times [I_-, I_+] \times \{\varphi \in \mathbb{T}\} \times \{s \in \mathbb{T}\} \times [-\varepsilon_0, \varepsilon_0]$$

to be the domain of definition of our problem. Hence, the $C^r$ norms of functions will refer to sup norms defined on this set. Of course, in the case that the functions depend only on a few of the variables (e.g. $V$, which only depends on $q$), we can also consider them as function of more variables and define the norm in the appropriate domain.

The main rigorous result of this paper is:

THEOREM 4.1. *Consider a Hamiltonian of the form (3.3) Assume:*

**H1**. *The terms $V$ and $h$ in (3.3) are uniformly $C^r$ for $r \geq r_0$, sufficiently large.*

**H2**. *The potential $V : \mathbb{T} \to \mathbb{R}$ has a unique global maximum at $q = 0$ which is non-degenerate (i.e., $V''(0) < 0$). We denote $(q_0(t), p_0(t))$ an orbit of the pendulum $P_\pm(p, q)$ in (3.2), homoclinic to $(0, 0)$.*

**H3**. *$h$ is a trigonometric polynomial in $\varphi$ and $s$:*

(4.2) $$h(p, q, I, \varphi, s; \varepsilon) = \sum_{k, l \in \mathcal{N}} \hat{h}_{k,l}(p, q, I; \varepsilon) e^{i(k\varphi + ls)},$$

*where $\mathcal{N} \subset \mathbb{Z}^2$ is a finite set.*

**H4**. *Consider the Poincaré function, also called Melnikov potential, associated to $h$ (and to the homoclinic orbit $(q_0, p_0)$ mentioned in **H2**):*

(4.3) $$\mathcal{L}(I, \varphi, s) = -\int_{-\infty}^{+\infty} \Big( h\left(p_0(\sigma), q_0(\sigma), I, \varphi + I\sigma, s + \sigma; 0\right)$$
$$- h(0, 0, I, \varphi + I\sigma, s + \sigma; 0) \Big) d\sigma.$$

*Assume that, for any value of $I \in (I_-, I_+)$ there exists a nonempty open set $\mathcal{J}_I \subset \mathbb{T}^2$, with the property that when $(I, \varphi, s) \in H_-$, where*

$$(4.4) \qquad H_- = \bigcup_{I \in (I_-, I_+)} \{I\} \times \mathcal{J}_I \subset (I_-, I_+) \times \mathbb{T}^2,$$

*the map*

$$(4.5) \qquad \tau \in \mathbb{R} \mapsto \mathcal{L}(I, \varphi - I\tau, s - \tau) \equiv \Gamma(\tau; I, \varphi, s)$$

*has a non-degenerate critical point $\tau$ which is locally given, by the implicit function theorem in the form $\tau = \tau^*(I, \varphi, s)$ with $\tau^*$ a smooth function.*

*Assume moreover that for every $(I, \varphi, s) \in H_-$, the function*

$$(4.6) \qquad \frac{\partial \mathcal{L}}{\partial \varphi}(I, \varphi - I\tau^*, s - \tau^*)$$

*is non-constant and negative (respectively positive).*

**H5**. *The perturbation terms $h(p, q, I, \varphi, s; 0)$, $\frac{\partial h}{\partial \varepsilon}(p, q, I, \varphi, s; 0)$ satisfy some non-degeneracy conditions (which will be stated explicitly on page 60 of Section 8.5.2, and in equation (10.9) on page 106, and equation (10.14) on page 112 of Section 10.2, as **H5'**, **H5"**, and **H5'''**, respectively).*

Then, there is $\varepsilon^* > 0$ such that for $0 < |\varepsilon| < \varepsilon_*$ and for any interval $[I_-^*, I_+^*] \subset (I_-, I_+)$, there exists a trajectory $\tilde{x}(t)$ of the system (3.4) such that for some $T > 0$

$$(4.7) \qquad I(\tilde{x}(0)) \leq I_-^*; \quad I(\tilde{x}(T)) \geq I_+^*$$

(respectively:

$$(4.8) \qquad I(\tilde{x}(0)) \geq I_+^*; \quad I(\tilde{x}(T)) \leq I_-^*).$$

We will consider $I_- < I_+$ as fixed and somewhat large. In particular, $[I_-, I_+]$ can contain all the resonances $I = -l/k$, for $(k, l) \in \mathcal{N}$. Then, the trajectories that we construct cross over the resonant regions. Hence, we overcome the *large gap problem* by showing the existence of orbits which traverse regions in which primary KAM tori are not present and indeed, there are no transverse heteroclinic intersections between them accessible to direct perturbation theory.

In this paper, we will not address rigorously the issue of how abundant is the mechanism presented here. Nevertheless, in Section 12.3, we will present some heuristic remarks on the abundance of systems satisfying the hypotheses of Theorem 4.1.

REMARK 4.2. Note that **H4** contains two hypotheses.

First, the existence of non-degenerate critical points for the map (4.5). This first hypothesis will imply that, for $0 < |\varepsilon| \ll 1$, the stable and unstable invariant manifolds of $\Lambda_\varepsilon$—which agreed for $\varepsilon = 0$—will have a transversal intersection in the domain that we are considering.

Second, that the function (4.6) has a sign. The sign assumed is linked to the fact that the trajectory provided by Theorem 4.1 increases its $I$ coordinate. For the experts we anticipate that the proof of Theorem 4.1 consists on constructing transition chains that start in $\tilde{x}(0)$ and end in $\tilde{x}(T)$.

We note that both conditions are open. If there is a non-degenerate critical point of the map (4.5) for some $(I, \varphi, s)$, in a neighborhood of these variables we can also find a locally unique non-degenerate critical point. If we consider the critical points produced by the implicit function theorem, we see that the negativity condition for the map (4.6) will also hold. Note that the domains where these properties hold are independent of $\varepsilon$.

As we will see, we will not need to assume that, for different $I$, we use critical points that are obtained by applying the implicit function theorem. Each of the critical points will produce heteroclinic connections among KAM tori whose $I$ covers an interval independent of $\varepsilon$. By choosing a finite number of these critical points, we will be able to transverse the region that covers the resonances.

REMARK 4.3. There are several variants of hypothesis **H4** that lead, with the methods of this paper, to other variants of the main result.

For instance, if we assume that there are different sets $H_-^+$, $H_-^-$ which satisfy the second part of hypothesis **H4** with positive and negative sign, respectively, and such that the range in the variable $I$ overlaps, we can find orbits which perform largely arbitrary excursions in $I$. The reason is that, if the $I$ projections of $H_-^+$, $H_-^-$ overlap, we can continue a transition chain in $H_-^+$ by a transition chain in $H_-^-$ and repeat, so that, along this whole chain $I$ increases and decreases.

One important case when there are intervals with both signs, overlapping in the variable $I$, is when the existence of non-degenerate critical points for the map (4.5) happens for all $(\varphi, s) \in \mathbb{T}^2$. (An explicit example of this situation is explained in detail in Chapter 13.) In such a case, if the Melnikov potential (4.3) is non constant, then $\dfrac{\partial \mathcal{L}}{\partial \varphi}$ has to have an interval where it is positive and another where it is negative. The reader may want to check this case first to gain intuition.

REMARK 4.4. In general, the pendulum $P_\pm$ will have two homoclinic orbits. To each of these homoclinic orbits we can associate a Poincaré function via (4.3). For generic perturbations, each of the different Poincaré functions will have several non-degenerate critical points. It can therefore happen that one can find several intervals $I^k \equiv [I_-^{*,k}, I_+^{*,k}]$ on which one can verify the hypotheses of Theorem 4.1. If there is a closed interval $I^* = [I_-^*, I_+^*]$ contained in the union of the interior of the intervals $I_k$, then, we also obtain the conclusions of Theorem 4.1 for the interval $I^*$. As we will see in more detail in Section 12.3 this makes it much easier to verify the existence of the mechanism in concrete models. The reason is that if we consider just one smooth curve $\tau^*$ it often happens that this manifold has codimension one boundary in which the critical point becomes degenerate. If the boundaries for different $\tau^*$'s do not coincide, the diffusion can proceed. Moreover, we will see that if the projections in $I$ of the region where the positive and negative alternatives of (4.6) hold, it is possible to find orbits whose $I$ performs largely arbitrary excursions.

REMARK 4.5. We have taken advantage of the symplectic geometry to express the sufficient condition of transversality of stable and unstable invariant manifolds in terms of only one function, called here Poincaré function or Melnikov potential. This is very natural geometrically since the stable and unstable manifolds of whiskered tori are Lagrangian. See [**DG00, DG01**]. As it turns out, the same Melnikov potential will appear in the conditions **H5**, a connection which is not apparent if one carries out the perturbation theory without using the symplectic structure.

REMARK 4.6. We also note that in [**CG94**] it is shown that in the model (3.3), under the hypothesis that the perturbation contains a finite number of harmonics—which we also include at the moment—one can obtain diffusion in the $I$ direction which is of a size independent of $\varepsilon$. The reason is that, outside the resonances $I = -l/k$, for $(k, l) \in \mathcal{N}$, $k \neq 0$, which are a finite set, they show that the primary KAM tori are very close. (This is called in [**CG94**] the *gap bridging* mechanism.) Note that what is shown in [**CG94**] is not that there are orbits that transverse the resonant gaps, but rather that, in the regions that contain no resonances there are no large gaps between primary KAM tori. The argument of [**CG94**] is completely different from the mechanism considered here. In our case, there are inded large gaps, but the mechanism introduced here overcomes them by using secondary tori and lower dimensional tori.

REMARK 4.7. In the paper [**Gal99**] one can find a treatment of diffusion along the action variables $A$. Since the KAM tori contain one-parameter families in the variable $A$ (see Remark 3.2), there are no gaps among them and it is possible to construct transition chains along them.

The phenomena discussed in this paper are very different from the phenomena established in [**Gal99**] since the diffusion we establish is present in the non-autonomous version of the problem which ignores the variable $A$. The non-autonomous system does present gaps among the KAM tori and, indeed, our transition chains have to include objects other than KAM tori.

REMARK 4.8. A value of $r_0$ which follows from our argument is $r_0 = 60$. Of course, this is not optimal even for the argument presented, and better regularities can presumably be obtained using different methods in part or in all of our argument. Clearly, using topological or variational methods in parts of the argument, or claiming generic results will require significantly lower regularity. Since the emphasis along this paper is on geometric objects and it is widely believed that diffusion is more difficult in the analytic case, we have not tried very hard to lower the differentiability requirements.

REMARK 4.9. We think that Hypothesis **H3** on the finiteness of the number of harmonics of the perturbation can presumably be eliminated in Theorem 4.1 at the price of complicating some of the estimates presented here. We will present some heuristic considerations in Section 12.4 but we will postpone now such considerations.

We note, however, that the hypothesis that is really used in this paper is slightly weaker than **H3**. We just need that there are a finite number of resonances, but the number of Fourier terms associated to each resonance could be infinite. Since the formulation of this more precise hypothesis is more technical, we have just used the simpler version of trigonometric polynomial.

REMARK 4.10. Thanks to the modularity of the method used in the proof of the main result, it can also be applied to situations in which $(I, \varphi)$ are higher dimensional or in which $s$ ranges in a torus or in which the oscillator is substituted by a rotator (an anisochronous system in the notation of [**Gal94**]). Analogously for the case of higher dimensional $(p, q)$. All these topics are currently being researched.

REMARK 4.11. Even if this paper is logically independent from [**DLS00**], the analysis here is inspired by that of [**DLS00**] and we will look for objects quite analogous to those considered in that paper. Indeed, we will keep many of the notations introduced there since this will make it easier to use some of the results of that paper in the argument here. Nevertheless, there will be quite significant differences between the way that they are organized. Roughly speaking, the fixed point of the pendulum in (3.3) will play the role that the periodic geodesic played in [**DLS00**]. One of the separatrices of the pendulum will play the role of the connecting geodesic in [**DLS00**].

In Section 12.2 we have undertaken a more systematic comparison between the objects and the methods of [**DLS00**] and this paper.

In the subsequent sections we will give a detailed proof of Theorem 4.1. In fact we will establish this Theorem by checking that system (3.3) satisfies the mechanisms explained in Section 2.2.

# CHAPTER 5

# Notation and definitions, resonances

We will say that a function satisfies $F = O_{C^r}(\eta)$ when

$$||F||_{C^r} \le \text{cte.}\,\eta.$$

Given a function depending on the $\varepsilon$ variable—and others—in a sufficiently smooth fashion, we will introduce the following notation to denote the partial Taylor expansion in $\varepsilon$

$$(5.1) \quad F(x;\varepsilon) = F_0(x) + \varepsilon F_1(x) + \varepsilon^2 F_2(x) + \cdots + \varepsilon^n F_n(x) + O(\varepsilon^{n+1}).$$

Note that if $F$ is $C^r$ in all its variables, including $\varepsilon$, the remainder will be $O_{C^{r-n-1}}(\varepsilon^{n+1})$ considered as a function of all the variables. This counting of derivatives is somewhat wasteful, but avoiding it will require to keep track of separate regularities.

Given a trigonometric polynomial in the angle variables $(\varphi, s)$

$$(5.2) \quad F(p,q,I,\varphi,s) = \sum_{(k,l)\in\mathcal{N}} \hat{F}_{k,l}(p,q,I) e^{i(k\varphi + ls)},$$

we denote by $\mathcal{N} = \mathcal{N}(F) \subset \mathbb{Z}^2$, the support of the Fourier transform.

It is obvious that:

$$\mathcal{N}(h + g) \subset \mathcal{N}(h) \cup \mathcal{N}(g)$$
$$\mathcal{N}(hg) \subset \mathcal{N}(h) + \mathcal{N}(g)$$
$$(5.3) \qquad \mathcal{N}(\nabla h) \subset \mathcal{N}(h)$$
$$\mathcal{N}\left(\int h\,dI\right) \subset \mathcal{N}(h)$$

where by the addition of subsets in $\mathbb{Z}^2$ we denote the set consisting of sums of an element of the first set and an element of the second.

For a function $F(p,q,I,\varphi,s;\varepsilon)$ of the form (5.2), which moreover depends on a parameter $\varepsilon$, the set

$$(5.4) \qquad \mathcal{R}_1 = \mathcal{R}_1(F) := \{I = -l/k : (k,l) \in \mathcal{N}(F_1), k \ne 0\}$$

will be called the set of *primary resonances*. As it will be developed in more detail later (and is well known to experts) this is the region to avoid in first order averaging. (See Section 8.3 for a description of the averaging method and a clarification of why the set $\mathcal{R}_1$ plays a role.)

The region

$$\begin{aligned}\mathcal{R}_2 &= \mathcal{R}_2(F)\\ &:= \{I = -l/k : (k,l) \in (\mathcal{N}(F_1) + \mathcal{N}(F_1)) \cup \mathcal{N}(F_2), k \neq 0\}\end{aligned}$$

(5.5)

will be called the set of *secondary resonances*. (Roughly speaking, $\mathcal{R}_2$ is the set where it is impossible to apply second order averaging.) These definitions can be generalized to the sets $\mathcal{R}_k$, of the resonances of order $k$ which are the places where the $k^{\text{th}}$ order averaging cannot be applied (see Section 8.3). The main resonant set for our model (3.3) will be

$$(5.6) \qquad\qquad\qquad \mathcal{R}_1(h) \cup \mathcal{R}_2(h).$$

As it turns out, in our problem, resonances of order higher than the second will play a very small role. When $h(p, q, I, \varphi, s; \varepsilon)$ is a trigonometric polynomial in $(\varphi, s)$, $\mathcal{R}_1(h) \cup \mathcal{R}_2(h)$ is a finite set.

The main step in our proof will be to jump over regions close to the main resonance set (5.6). As we will show later in detail in Section 8.3, outside of the main resonant region, the KAM tori are close enough so there are no large gaps between them and the mechanism in [**Arn64**] applies.

# CHAPTER 6

# Geometric features of the unperturbed problem

In this chapter we will discuss the geometric features of the unperturbed system which will survive under the perturbation. They will serve as landmarks to organize the motion of the perturbed system.

The main feature of Hamiltonian (3.3) for $\varepsilon = 0$ is that it consists of two uncoupled systems (a rotator and a pendulum) so that the Cartesian product of invariant objects of each of the subsystems will give an invariant object of the full system.

To simplify the notation and without loss of generality, we have chosen coordinates in **H2** in such a way that the maximum of $V$ happens at $q = 0$. Then, for the pendulum described by the Hamiltonian $P_\pm(p,q)$, the point $p = 0$, $q = 0$ is a fixed point and is hyperbolic as an orbit in $\mathbb{R} \times \mathbb{T}$, the phase space of the pendulum.

The stable and unstable manifolds of this point coincide. We will denote by

$$(6.1) \qquad\qquad \gamma(t) := (p_0(t), q_0(t))$$

one of the orbits of $P_\pm$ which is homoclinic to the hyperbolic point $(0,0)$.

REMARK 6.1. As it is well known in mechanics, the case that $P_\pm(p,q) = \pm(p^2/2 + \cos q - 1)$ is the physical pendulum, the parameterizations of the separatrices have explicit formulas given by

$$(6.2) \qquad\qquad q_0(t) = 4 \arctan e^{\pm t}, \quad p_0(t) = 2/\cosh t.$$

and

$$(6.3) \qquad\qquad q_0(t) = 4 \arctan e^{\mp t}, \quad p_0(t) = -2/\cosh t.$$

In Chapter 13, we will present many other explicit calculations for the case of the pendulum.

When we consider the full system (3.4) for $\varepsilon = 0$, we see that, for any value of $I \in \mathbb{R}$, the product of the hyperbolic point of the pendulum with all the other variables will be a 2-dimensional invariant torus in the extended phase space

$$\mathcal{T}_I = \{\tilde{x} = (p, q, I, \varphi, s) : p = q = 0, \ (\varphi, s) \in \mathbb{T}^2\},$$

and one component of the stable and unstable manifolds of this torus, $W^s_{\mathcal{T}_I}$, $W^u_{\mathcal{T}_I}$, coincides along the 3-dimensional manifold given by:

$$\{(p_0(\tau), q_0(\tau), I, \varphi, s) : \tau \in \mathbb{R}, \ (\varphi, s) \in \mathbb{T}^2\},$$

where $(p_0(\tau), q_0(\tau))$ is given in (6.1).

Hence, for the values $I_-, I_+$, given in Theorem 4.1, we introduce the set

$$\tilde{\Lambda} = \{\tilde{x} \in (\mathbb{R} \times \mathbb{T})^2 \times \mathbb{T} : p = q = 0, I \in [I_-, I_+]\}$$
(6.4)
$$= \bigcup_{I \in [I_-, I_+]} \mathcal{T}_I$$

The set $\tilde{\Lambda}$ is a 3-dimensional manifold with boundary. It is locally invariant and normally hyperbolic for the flow (3.4) for $\varepsilon = 0$. The manifold $\tilde{\Lambda}$ is topologically $[I_-, I_+] \times \mathbb{T}^2$. It has a global system of coordinates given by $(I, \varphi, s)$. Hence, we will not need to distinguish very explicitly whether we are talking about the three numbers $(I, \varphi, s)$ or the point in the manifold $\tilde{\Lambda}$. Nevertheless, we will emphasize as much as possible the geometric meaning of the calculations that need to be carried out in coordinates.

The unperturbed flow on $\tilde{\Lambda}$ is given by the quasi-periodic flow:

(6.5)        $\Phi_{t,0}(0, 0, I_0, \varphi_0, s_0) = (0, 0, I_0, \varphi_0 + I_0 t, s_0 + t),$

and we will be denote by

$$\tilde{\lambda}_t(I_0, \varphi_0, s_0) = (0, 0, I_0, \varphi_0 + I_0 t, s_0 + t)$$

the orbits in the torus $\mathcal{T}_{I_0}$.

One component of the stable and unstable invariant manifolds for $\tilde{\Lambda}$, $W^s_{\tilde{\Lambda}}$, $W^u_{\tilde{\Lambda}}$ coincides in a manifold $\tilde{\gamma}$ of orbits homoclinic to $\tilde{\Lambda}$

$$\tilde{\gamma} \subset (W^s_{\tilde{\Lambda}} \setminus \tilde{\Lambda}) \cap (W^u_{\tilde{\Lambda}} \setminus \tilde{\Lambda}).$$

and a parameterization of $\tilde{\gamma}$ is given by:

(6.6)     $\tilde{\gamma} := \{(p_0(\tau), q_0(\tau), I, \varphi, s) \ : \ I \in [I_-, I_+], \ \tau \in \mathbb{R}, \ (\varphi, s) \in \mathbb{T}^2\},$

where $(p_0, q_0)$ is, as in (6.1), the chosen homoclinic orbit of the pendulum.

Hence, the meaning of the coordinate $\tau$ is the time of the flow along the unperturbed separatrix. We denote by

$$\begin{aligned}\tilde{\gamma}_t(\tau, I_0, \varphi_0, s_0) &= (p_0(\tau + t), q_0(\tau + t), I_0, \varphi_0 + I_0 t, s_0 + t)\\ &= \Phi_{t,0}(p_0(\tau), q_0(\tau), I_0, \varphi_0, s_0),\end{aligned}$$

the unperturbed flow on $\tilde{\gamma}$, our chosen component of the homoclinic manifold.

We note that for any $\tau \in \mathbb{R}$

(6.7)     $\text{dist}(\tilde{\lambda}_t(I_0, \varphi_0, s_0), \tilde{\gamma}_t(\tau, I_0, \varphi_0, s_0)) \to 0 \quad \text{for} \quad t \to \pm\infty.$

Let us note that the unperturbed system is a product, and that $(0, 0)$ is a critical point of the pendulum with characteristic exponents $\mu_\pm = \pm\mu$, where $\mu := (-V''(0))^{1/2}$. Moreover, the exponents of contraction in the tangent direction of $\tilde{\Lambda}$ are 0 (see (6.5)). Then, the stable and unstable manifolds of $\tilde{\Lambda}$ are characterized as the set of orbits whose distance to the orbits in $\tilde{\Lambda}$ is less than $C \exp(-\mu |t|)$ respectively as $t \to \pm\infty$, and we have

$$\mathrm{dist}(\tilde{\lambda}_t(I_0, \varphi_0, s_0), \tilde{\gamma}_t(\tau, I_0, \varphi_0, s_0)) \leq C(\tau) \exp(-\mu |t|) \quad \text{for} \quad t \to \pm\infty.$$

CHAPTER 7

# Persistence of the normally hyperbolic invariant manifold and its stable and unstable manifolds

Since the manifold $\tilde{\Lambda}$ is normally hyperbolic, and is locally invariant for the flow (3.4) for $\varepsilon = 0$, by the theory of normally hyperbolic manifolds we have that the manifold persists under small perturbations. Moreover, if the system depends smoothly on parameters, the manifolds—they may be non unique—may be chosen to depend smoothly on parameters.

A formulation of the results of [**Fen72, Fen74, Fen77**] in the way that we will use them (very similar to the statement of Theorem 4.2 of [**DLS00**]. See also appendix A of [**DLS05**], which contains a detailed proof of an slightly more general result) is:

THEOREM 7.1. *Consider a Hamiltonian as in* (3.3). *Assume that $H_\varepsilon$ is uniformly $C^r$, $r \geq 2$ in all its variables, including $\varepsilon$, in a neighborhood of $\tilde{\Lambda}$ and $\tilde{\gamma}$.*

*Then, there exists $\varepsilon^* > 0$ such that for $|\varepsilon| < \varepsilon^*$, there is a $C^{r-1}$ function*

$$\tilde{\mathcal{F}} : \tilde{\Lambda} \times (-\varepsilon^*, \varepsilon^*) \longrightarrow (\mathbb{R} \times \mathbb{T})^2 \times \mathbb{T}$$

*such that*

(7.1) $$\tilde{\Lambda}_\varepsilon := \tilde{\mathcal{F}}(\tilde{\Lambda} \times \{\varepsilon\})$$

*is locally invariant for the flow* (3.4) *generated by the vector field $\mathcal{H}_\varepsilon$.*

*In particular, $\tilde{\Lambda}_\varepsilon$ is $\varepsilon$-close to $\tilde{\Lambda}_0 = \tilde{\Lambda}$ in the $C^{r-2}$ sense.*

*Moreover, $\tilde{\Lambda}_\varepsilon$ and the vector field $\mathcal{H}_\varepsilon$ can be extended so that it is a normally hyperbolic invariant manifold for the flow* (3.4) *generated by $\mathcal{H}_\varepsilon$. In particular, it is possible to define local stable and unstable manifolds. That is, we can find a $C^{r-1}$ function $\tilde{\mathcal{F}}^{\mathrm{s}}$ such that the (local) stable manifold of $\tilde{\Lambda}_\varepsilon$ takes the form*

(7.2) $$W^{\mathrm{s,loc}}_{\tilde{\Lambda}_\varepsilon} = \tilde{\mathcal{F}}^{\mathrm{s}}\left(\tilde{\Lambda} \times (0, +\infty) \times \{\varepsilon\}\right).$$

*If $\tilde{x} = \tilde{\mathcal{F}}(I, \varphi, s; \varepsilon) \in \tilde{\Lambda}_\varepsilon$, then $W^{\mathrm{s,loc}}_{\tilde{x}} = \tilde{\mathcal{F}}^{\mathrm{s}}(\{I\} \times \{\varphi\} \times \{s\} \times (0, \infty) \times \{\varepsilon\})$. In particular, $W^{\mathrm{s,loc}}_{\tilde{\Lambda}_\varepsilon}$ is $\varepsilon$-close to $W^{\mathrm{s,loc}}_{\tilde{\Lambda}}$ in the $C^{r-2}$ sense.*

*Analogous results hold for the (local) unstable manifold $W^{\mathrm{u,loc}}_{\tilde{\Lambda}_\varepsilon}$.*

The proof of Theorem 7.1 is quite standard in the theory of normally hyperbolic invariant manifolds. It is a straightforward application of the theorems in [**Fen74, Fen77, HPS77**]. A modern proof is in [**Lla00**]. A detailed proof tailored for the models considered here is in Appendix A of [**DLS05**].

A useful observation is that there is an easy way to obtain smooth dependence on parameters from the standard results on persistence. It suffices to consider the system obtained by taking the product of the original system and the identity in the direction of parameters. The fact that the invariant manifolds for the extended system are regular, gives the fact that the manifolds of the original system depend regularly on parameters.

Note that the coordinates along the unperturbed manifold $\tilde{\Lambda}$ under the unperturbed evolution just rotate or remain invariant. Since $||D\Phi_{t,0}|_{\tilde{\Lambda}}|| \leq C|t|$ (see (6.5)), we have that for every $\delta > 0$,

$$||\Phi_{t,0}|_{\tilde{\Lambda}}|| \leq C_\delta e^{\delta|t|}.$$

This shows that the tangential exponents along the manifold can be taken as small as we want.

On the other hand, because of assumption **H2**, the point $(p,q) = (0,0)$ is a hyperbolic point for the pendulum. Since the system (3.4) for $\varepsilon = 0$ is a direct product of rotators—along $\tilde{\Lambda}$—and a pendulum, the hyperbolic directions of the pendulum become the stable/unstable bundles of the manifold $\tilde{\Lambda}$.

The fact that the tangential exponents are arbitrarily small allows us to conclude that the manifold is as regular as the flow when measured on the $C^{r-1}$ classes, $r \in \mathbb{N}$. (If there was a non-trivial expansion exponent, then the regularity claimed for the manifold would be the infimum of $r-1$ and a limiting regularity determined by the rates of expansion along the manifold and along the stable/unstable bundles.)

REMARK 7.2. In the general theory of the persistence of overflowing locally invariant manifolds, the manifold obtained need not be unique since, in principle, it could depend on some of the choices made in the proof. Nevertheless, when the manifold $\tilde{\Lambda}_\varepsilon$ is invariant and not just locally invariant, it is unique. We will show later that in the manifold $\tilde{\Lambda}_\varepsilon$ there are codimension 1 KAM tori. Therefore we conclude that the portion of $\tilde{\Lambda}_\varepsilon$ between the KAM tori is actually invariant. Hence, for our case, the manifolds will turn out to be unique. Nevertheless, we will not take advantage of this fact since it does not simplify appreciably the proofs and it can fail in more complicated models that can be analyzed by the methods presented here.

REMARK 7.3. Note that, as we discuss after Theorem 7.1, the parameterization $\tilde{\mathcal{F}}$ is not uniquely defined even after the locally invariant manifold has been fixed. Indeed, if we compose $\tilde{\mathcal{F}}$ with any diffeomorphism of $\tilde{\Lambda}$ on the right we obtain a different parameterization.

We can make the parameterization unique by imposing that $\tilde{\mathcal{F}}$ does not change the $(I, \varphi, s)$ coordinates:

(7.3)     $\Pi_I \tilde{\mathcal{F}}(I, \varphi, s; \varepsilon) = I, \ \Pi_\varphi \tilde{\mathcal{F}}(I, \varphi, s; \varepsilon) = \varphi, \ \Pi_s \tilde{\mathcal{F}}(I, \varphi, s; \varepsilon) = s.$

This can be expressed as saying that we write $\tilde{\Lambda}_\varepsilon$ as the graph of a function from $\tilde{\Lambda}$ into the $(p, q)$ plane.

This normalization will be important for us when we compute the map $\tilde{\mathcal{F}}$ in coordinates.

REMARK 7.4. We recall that the precise definition of the stable manifolds in normal hyperbolicity theory is that they consist on the points that, under the evolution of the extended system, converge to the invariant manifold with an exponential rate close to that of the original system. It follows from the exponential convergence that $W^{\mathrm{s}}_{\tilde{\Lambda}_\varepsilon} = \cup_{\tilde{x} \in \tilde{\Lambda}_\varepsilon} W^{\mathrm{s}}_{\tilde{x}}$ where

$$W^{\mathrm{s}}_{\tilde{x}} = \{\tilde{y} \ : \ \mathrm{dist}(\Phi_{t,\varepsilon}(\tilde{x}), \Phi_{t,\varepsilon}(\tilde{y})) \leq C e^{-\tilde{\mu} t}, t > 0\},$$

where $\tilde{\mu} = \mu + \mathrm{O}(\varepsilon)$. Moreover $W^{\mathrm{s}}_{\tilde{x}} \cap W^{\mathrm{s}}_{\tilde{y}} = \emptyset$ when $\tilde{x} \neq \tilde{y}$, $\tilde{x}, \tilde{y} \in \tilde{\Lambda}_\varepsilon$. The stable manifolds of the points are as smooth as the flow (when smoothness is measured in the class of $C^r$ spaces $r \in \mathbb{N} \cup \{\infty, \omega\}$). Nevertheless, the map $\tilde{x} \mapsto W^{\mathrm{s}}_{\tilde{x}}$ could be less regular than the flow, and therefore, the total manifold $W^{\mathrm{s}}_{\tilde{\Lambda}_\varepsilon} = \cup_{\tilde{x} \in \tilde{\Lambda}_\varepsilon} W^{\mathrm{s}}_{\tilde{x}}$ could, in general, be less smooth than the flow. In our case, however, since the tangent exponents are arbitrarily small, if the flow is $C^{r-1}$, $r - 1 \in \mathbb{N}$, then the resulting $W^{\mathrm{s}}_{\tilde{\Lambda}_\varepsilon}$ will be $C^{r-1}$ for small $\varepsilon$.

Analogous results hold for the unstable manifold.

Note that the definition of the (un)stable manifolds for locally invariant manifolds can only be made in an extended system constructed in the proof for which the dynamics is defined for all times. In general, these (un)stable manifolds depend on the extended flow used.

One consequence of that is that $\tilde{\Lambda}_\varepsilon$, $W^{\mathrm{s}}_{\tilde{\Lambda}_\varepsilon}$ may fail to be $C^\infty$ even if the flow is analytic. Nevertheless, when the manifolds are invariant, the stable and unstable manifolds are uniquely defined. In our case, as stated in Remark 7.2, the KAM tori will produce invariant boundaries for $\tilde{\Lambda}_\varepsilon$, hence stable and unstable manifolds will be uniquely defined. The proofs we present, however, do not take advantage of this fact.

## 7.1. Explicit calculations of the perturbed invariant manifold

In the rest of this section, we will compute rather explicitly the expansions in $\varepsilon$ of the map $\tilde{\mathcal{F}}$ for the Hamiltonian (3.3).

We note that the equation that $\tilde{\mathcal{F}}$ has to satisfy so that its range is invariant by the vector field $\mathcal{H}_\varepsilon$ given in (3.4) is

$$(7.4) \qquad\qquad \mathcal{H}_\varepsilon \circ \tilde{\mathcal{F}} = D\tilde{\mathcal{F}} \mathcal{R}$$

where $\mathcal{R}$ is a vector field in $\tilde{\Lambda}_\varepsilon$. Note that (7.4) expresses that the vector field $\mathcal{H}_\varepsilon$ at a point in the range is tangent to the range of $\tilde{\mathcal{F}}$.

Using equation (7.4) one can find an expansion of the function $\tilde{\mathcal{F}}$.

PROPOSITION 7.5. *The family of mappings $\tilde{\mathcal{F}}$ specified in Theorem 7.1 with the normalization (7.3) admits an expansion*

$$\tilde{\mathcal{F}} = \tilde{\mathcal{F}}_0 + \varepsilon\tilde{\mathcal{F}}_1 + \cdots + \varepsilon^m\tilde{\mathcal{F}}_m + \mathrm{O}_{C^{r-m-2}}(\varepsilon^{m+1}),$$

*where $\tilde{\mathcal{F}}_0(I, \varphi, s) = (0, 0, I, \varphi, s)$.*

*In the case that the flow satisfies assumptions* **H1**–**H3**, *then the functions $\tilde{\mathcal{F}}_1, \ldots, \tilde{\mathcal{F}}_m$ are trigonometric polynomials in $\varphi, s$, and $\tilde{\mathcal{F}}_i$ are of class $C^{r-1-i}$. Moreover, in such a case, $\mathcal{N}(\tilde{\mathcal{F}}_1) \subset \mathcal{N}(h_1)$, $\mathcal{N}(\tilde{\mathcal{F}}_2) \subset (\mathcal{N}(h_1) + \mathcal{N}(h_1)) \cup \mathcal{N}(h_2)$.*

PROOF. Once we impose normalization (7.3) to $\tilde{\mathcal{F}}$, we can compute $\tilde{\mathcal{F}}_i$ by matching powers of $\varepsilon$ in the equation (7.4). We know by Theorem 7.1 that the expansion exists.

Equating terms in the expansion on $\varepsilon$ of the equation for invariance (7.4) we obtain, up to order two:

$$\mathcal{H}_0 \circ \tilde{\mathcal{F}}_0 = D\tilde{\mathcal{F}}_0\mathcal{R}_0$$

$$(D\mathcal{H}_0 \circ \tilde{\mathcal{F}}_0)\tilde{\mathcal{F}}_1 + \mathcal{H}_1 \circ \tilde{\mathcal{F}}_0 = D\tilde{\mathcal{F}}_0\mathcal{R}_1 + D\tilde{\mathcal{F}}_1\mathcal{R}_0$$

(7.5)
$$(D\mathcal{H}_0 \circ \tilde{\mathcal{F}}_0)\tilde{\mathcal{F}}_2 + \frac{1}{2}(D^2\mathcal{H}_0 \circ \tilde{\mathcal{F}}_0)\tilde{\mathcal{F}}_1^{\otimes 2} + (D\mathcal{H}_1 \circ \tilde{\mathcal{F}}_0)\tilde{\mathcal{F}}_1 + \mathcal{H}_2 \circ \tilde{\mathcal{F}}_0$$

$$= D\tilde{\mathcal{F}}_0\mathcal{R}_2 + D\tilde{\mathcal{F}}_1\mathcal{R}_1 + D\tilde{\mathcal{F}}_2\mathcal{R}_0.$$

In general, the equation obtained after matching the coefficients of $\varepsilon^n$ on both sides of (7.4), is of the form:

(7.6)    $$(D\mathcal{H}_0 \circ \tilde{\mathcal{F}}_0)\tilde{\mathcal{F}}_n - D\tilde{\mathcal{F}}_n\mathcal{R}_0 - D\tilde{\mathcal{F}}_0\mathcal{R}_n = -\mathcal{H}_n \circ \tilde{\mathcal{F}}_0 + \mathcal{S}_n$$

where $\mathcal{S}_n$ is polynomial in $\mathcal{H}_0, \ldots, \mathcal{H}_{n-1}$, their derivatives, $\tilde{\mathcal{F}}_0, \ldots, \tilde{\mathcal{F}}_{n-1}$, and $\mathcal{R}_0, \ldots, \mathcal{R}_{n-1}$.

Clearly, the first equation in (7.5) has as solution $\tilde{\mathcal{F}}_0 = (0, 0, I, \varphi, s)$, and $\mathcal{R}_0 = (0, I, 1)$. Therefore, if one can develop a method to solve equations for $\tilde{\mathcal{F}}_n$, $\mathcal{R}_n$ of the form of the linear Hamiltonian System (7.6) equal to a pre-assigned right hand side, we can keep on solving the hierarchy of equations (7.6) to any order.

In our case, since the unperturbed motion due to $\mathcal{R}_0$ is quasi-periodic, the equations of the form (7.6) can be solved quite explicitly using Fourier coefficients. (There are more general theories [**Lla00**] that allow to solve equations of the form (7.6) even if the motion on the base is not quasi-periodic.)

The theory of the equations that we need is summarized in the following Lemma 7.6. Clearly, applying it recursively, we obtain a proof of Proposition 7.5.                                                    □

LEMMA 7.6. *Let $\tilde{\mathcal{F}}_0$, $\mathcal{H}_0$, $\mathcal{R}_0$ be as in Proposition 7.5. Given a $C^s$ function, $s \geq 1$, $\eta : \tilde{\Lambda} \to \mathbb{R}^5$, we can find unique $C^1$ functions $\xi$, $\rho : \tilde{\Lambda} \to \mathbb{R}^5$ such that*

(7.7)    $$(D\mathcal{H}_0 \circ \tilde{\mathcal{F}}_0)\xi - D\xi\mathcal{R}_0 - D\tilde{\mathcal{F}}_0\rho = \eta$$

*and*

(7.8) $$\Pi_I \xi = 0, \ \Pi_\varphi \xi = 0, \ \Pi_s \xi = 0.$$

*Furthermore, $\xi$, $\rho$ are of class $C^s$ and we have, for a constant $C$ that depends only on $\mathcal{H}_0, \mathcal{R}_0$,*

$$||\xi||_{C^s} \leq C||\eta||_{C^s}$$
$$||\rho||_{C^s} \leq C||\eta||_{C^s}.$$

*Moreover, if $\eta$ is a trigonometric polynomial, so are $\xi$ and $\rho$, and we have $\mathcal{N}(\xi) \subset \mathcal{N}(\eta)$, $\mathcal{N}(\rho) \subset \mathcal{N}(\eta)$.*

PROOF. Note that the unperturbed flow and its differential at $\tilde{\Lambda}$ preserve the $I, \varphi, s$ directions. The plane $p, q$ is invariant.

Moreover, $D\tilde{\mathcal{F}}_0$ in the coordinates we are using is a $5 \times 3$ matrix. It consists of a $3 \times 3$ identity along the $I, \varphi, s$ directions and 0 along the other directions $(p, q)$.

The two observations above, immediately give us that we can satisfy the normalization (7.8) in a unique way by setting:

(7.9) $$\rho = \Pi_{I,\varphi,s}\eta$$

from which the regularity claims about $\rho$ follow.

Using again the invariance of the $(p, q)$ plane it suffices to study the equation

(7.10) $$(D\mathcal{H}_0 \circ \tilde{\mathcal{F}}_0)|_{p,q}\Pi_{p,q}\xi - D\Pi_{p,q}\xi\mathcal{R}_0 = \Pi_{p,q}\eta$$

This equation (7.10) can be further reduced by noticing that $D\mathcal{H}_0$ preserves the two eigendirections of the equilibrium point of the pendulum. Hence, if we denote by $\Pi^s$, $\Pi^u$, the projections along the stable and the unstable components and by $\mu$ the eigenvalue, we obtain that (7.10) is equivalent to:

(7.11) $$\mu\Pi^u\xi - D\Pi^u\xi\mathcal{R}_0 \ \ = \Pi^u\eta$$
$$-\mu\Pi^s\xi - D\Pi^s\xi\mathcal{R}_0 \ \ = \Pi^s\eta$$

The equations (7.11) can be studied easily. For example, they can be studied noting that the operator on the linear Hamiltonian system of (7.11) is diagonal on Fourier series. Hence, if

$$\Pi^s\eta(I, \varphi, s) = \sum_{(k,l)\in\mathcal{N}(\eta)} \hat{\eta}_{k,l}(I)e^{i(k\varphi+ls)},$$

the solution of (7.11) is:

(7.12) $$\Pi^s\xi(I, \varphi, s) = \sum_{(k,l)\in\mathcal{N}(\eta)} \hat{\eta}_{k,l}(I)[-\mu + i(kI + l)]^{-1}e^{i(k\varphi+ls)}$$

and analogously (with $+\mu$ in place of $-\mu$) for $\Pi^u\xi$.

Note that, since $\pm\mu \neq 0$, there are no small denominators in the finite sum (7.12). Then, one can bound the norm of $\xi$ by the norm of $\eta$. This finishes the proof of Lemma 7.6. $\square$

REMARK 7.7. Even if for us, the solution (7.12) is useful because it gives the Fourier coefficients that will later be the base of the study of resonances, we note that, in geometric perturbation theory, it is common to represent the solution as integral formulas.

For example, in our case, we have:

$$\Pi^s \xi(I, \varphi, s) = \int_0^\infty \eta(I, \varphi - It, s - t)e^{-\mu t}\, dt$$

Out of the preceding formula, it is easy to read off the regularity properties even in the case that $\eta$ is not a polynomial.

# CHAPTER 8

# The dynamics in $\tilde{\Lambda}_\varepsilon$

In this chapter, we will study the dynamics restricted to the manifold $\tilde{\Lambda}_\varepsilon$.

The upshot of the discussion will be that, under hypothesis of differentiability and **H5**, the invariant manifold $\tilde{\Lambda}_\varepsilon$ contains primary tori (see Propositions 8.21, 8.24, Theorem 8.30, Corollary 8.31), secondary tori (see Theorem 8.30, Corollary 8.31) and stable and unstable manifolds of periodic orbits (see Proposition 8.40). They are arranged in such a way that the gaps among them are a power of $\varepsilon$ which can be made as large as desired by assuming enough differentiability. For subsequent developments, any power greater than 1 will be enough. Hence, we have given the numerical values needed to obtain only gaps of order $\varepsilon^{3/2}$.

The results are summarized in Figure 8.1, which depicts a surface of section of the manifold $\tilde{\Lambda}_\varepsilon$.

The method of proof will be a sequence of different steps. Basically, they will be a combination of averaging methods and KAM theorems.

In Section 8.1 we study the geometry of the motion restricted to the manifold $\tilde{\Lambda}_\varepsilon$. In particular, we will see that the flow restricted to $\tilde{\Lambda}_\varepsilon$ is Hamiltonian. Moreover, for convenience, we will introduce a system of coordinates $(J, \varphi, s)$, in which the symplectic form has the standard expression.

In Section 8.2 we compute very explicitly the reduced Hamiltonian

$$k(J, \varphi, s; \varepsilon) = k_0(J, \varphi, s) + \varepsilon k_1(J, \varphi, s) + \cdots + \varepsilon^m k_m(J, \varphi, s) + \mathrm{O}(\varepsilon^{m+1})$$

which is the restriction of the Hamiltonian $\tilde{H}_\varepsilon$ to $\tilde{\Lambda}_\varepsilon$ expressed in the action-angle coordinates we have chosen in the invariant manifold. In particular, we will show that if the perturbation $h = h_1 + \varepsilon h_2 + \cdots + \varepsilon^{m-1} h_m + \mathrm{O}(\varepsilon^m)$ in (3.3) is a trigonometric polynomial with respect to $(\varphi, s)$, so are $k_0, k_1, \ldots, k_m$, and we will give rather explicit formulas to compute the $k_i$'s in terms of the $h_i$'s.

In Section 8.3 we develop a global averaging procedure that casts the reduced Hamiltonian $k(J, \varphi, s; \varepsilon)$ into a global normal form $\bar{k}(\mathcal{B}, \alpha, s; \varepsilon)$ which is given by different formulas in the resonant regions and in the non-resonant region.

The non-resonant region is studied in Section 8.4. Since the normal form $\bar{k}(\mathcal{B}, \alpha, s; \varepsilon)$ is very close to a strongly integrable Hamiltonian, a quantitative version of the KAM theorem, which we will develop, will show that the non-resonant region contains KAM tori which leave very small gaps between them.

The resonant region is analyzed in Section 8.5. In this region the normal form $\bar{k}(\mathcal{B}, \alpha, s; \varepsilon)$ is very close to a pendulum. Note that the pendulum has rotational and librational motions covering open sets as well as separatrices. The rotations in the pendulum have the same topology as the primary tori in the integrable system. The librations are contractible to a periodic orbit. Hence the librations correspond to motions with topologies that are not present in the integrable system.

The heart of the matter is Section 8.5.3 which shows that many of these rotational and librational orbits of the pendulum persist when we consider the error terms of the normal form. The rotational orbits will become primary KAM tori and the librational orbits will become secondary tori. For our purposes, it will be important to show that both the secondary tori and the primary tori can be found up to distances which are very close to the separatrices.

An important technical tool for the proofs of persistence of tori close to separatrices appears in Section 8.5.4 where we develop a system of action-angle coordinates in a neighborhood of the separatrices. This change of coordinates is singular (the $C^r$ norms blow up as a negative power of the distance to the separatrix). Since the remainder contains a power of $\varepsilon$, we will see that a KAM theorem can be applied provided that the distance to the separatrix is bigger than a constant power of $\varepsilon$. This power is arbitrarily high if the original system is differentiable enough.

In the language which is customary in the heuristic study of diffusion, what we have done is to develop upper bounds for the size of the chaotic sea. Related arguments for analytic mappings appear in [Neĭ84]. In the analytic case, the distance can be chosen to be exponentially small.

Furthermore, in Section 8.5.5, we study what happens to the separatrices of the pendulum when we include the error terms of the normal form. They become stable and unstable manifolds or lower dimensional tori (indeed, periodic orbits in our setting). See Figure 8.1 for a depiction of these invariant objects.

The distances between these invariant objects can be bounded thanks to the non-degeneracy assumption **H5** by a power of the perturbation. This power can be made arbitrarily large by assuming more differentiability. We will take it to be just $O(\varepsilon^{3/2})$ since this is enough for subsequent applications.

REMARK 8.1. Even if hypothesis **H5** is formulated in **H5'**, **H5"** and **H5"'** (on page 60 of Section 8.5.2 and in equations (10.9) on page 106 and (10.14) on page 112 of Section 10.2) in terms of the reduced Hamiltonian $k(J, \varphi, s; \varepsilon)$, we will have explicit formulas (see (8.5), (8.6)) that give the restricted Hamiltonian in terms of $H_\varepsilon$, the original one.

We have striven to make conditions **H5'**, **H5"** and **H5"'** quite explicit so that they can be verified rather easily for a concrete system. Nevertheless, we point out that the fact that they hold for a generic set—indeed a set of finite codimension—is rather easy.

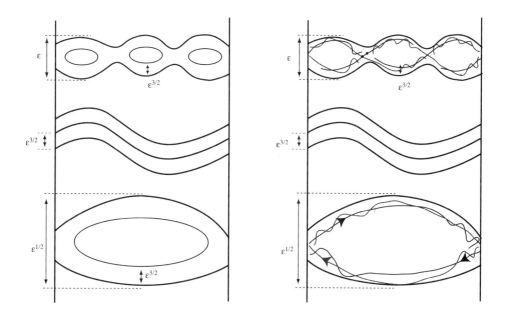

FIGURE 8.1. Surface of section of $\tilde{\Lambda}_\varepsilon$ illustrating the main invariant objects. The primary KAM tori and the secondary tori on the left. The primary tori and the stable and unstable manifolds of periodic orbits on the right.

## 8.1. A system of coordinates for $\tilde{\Lambda}_\varepsilon$

In this section, we will describe the construction of a system of coordinates in $\tilde{\Lambda}_\varepsilon$ that will help us to analyze the motion of the flow restricted to $\tilde{\Lambda}_\varepsilon$. In this system of coordinates, the symplectic form on $\tilde{\Lambda}_\varepsilon$ has the standard form and the flow is a Hamiltonian flow described by a time periodic one degree of freedom Hamiltonian.

We note that several of the notions that we use (e.g. resonances, which are related to the support of the Fourier transform) are dependent on the system of coordinates we use. Of course, we will check at the end that the geometric constructions are independent of the system of coordinates, but several of the analytic arguments using Fourier series, etc., depend on taking a system of coordinates.

Theorem 7.1 does not use much the structure of the flow (the only thing that we use of the structure of (3.3) is that $\|D\Phi_{t,0}|_{\tilde{\Lambda}}\| \leq$ cte. $t$). In the following paragraphs, we will develop some consequences of the facts that the perturbed and unperturbed flow are the autonomization of a Hamiltonian periodic flow.

The fact that the coordinate $s$ evolves always according to $\dot{s} = 1$, implies that we can decompose the manifold $\tilde{\Lambda}_\varepsilon$ as

$$\tilde{\Lambda}_\varepsilon = \cup_{s \in \mathbb{T}^1} \Lambda_\varepsilon^s \times \{s\}$$

according to the values that the coordinate $s$ takes.

PROPOSITION 8.2. *If $r \geq 3$, in the manifold $\tilde{\Lambda}_\varepsilon$ there is a $\mathcal{C}^{r-2}$ system of coordinates $(J, \varphi, s)$, where $J = \mathcal{J}(I, \varphi, s; \varepsilon)$, which is characterized uniquely by the following conditions:*

   i) *The angle coordinates $\varphi$ and $s$ are the same as those of the unperturbed system.*
   ii) *The symplectic form in $\Lambda_\varepsilon^s$ is given by $\omega|_{\Lambda_\varepsilon^s} = dJ \wedge d\varphi$.*
   iii) *$\mathcal{J}(0, \varphi, s; \varepsilon) = 0$.*

*Moreover:*

   iv) *The function $\mathcal{J}(I, \varphi, s; \varepsilon)$ is $\mathcal{C}^{r-2}$ and is $\varepsilon^2$-close, in the $\mathcal{C}^{r-3}$ sense, to the function $I$.*
   v) *There are expansions:*

$$\mathcal{J}(I, \varphi, s; \varepsilon) = I + \varepsilon^2 \mathcal{J}_2(I, \varphi, s) + \cdots + \varepsilon^m \mathcal{J}_m(I, \varphi, s) + O_{\mathcal{C}^{r-m-3}}(\varepsilon^{m+1}),$$

   *where $\mathcal{J}_i$, of class $\mathcal{C}^{r-i-1}$, are trigonometric polynomials in $(\varphi, s)$. Moreover $\mathcal{N}(\mathcal{J}_2) \subset \mathcal{N}(h_1) + \mathcal{N}(h_1)$.*

PROOF. It is clear that $\Phi_{t,\varepsilon}(\Lambda_\varepsilon^s \times \{s\}) = \Lambda_\varepsilon^{s+t} \times \{s+t\}$.

Since $\Lambda_0^s = \Lambda := \{(0, 0, I, \varphi) : (I, \varphi) \in [I_-, I_+] \times \mathbb{T}\}$ for any $s \in \mathbb{T}$, we have that $\omega|_{\Lambda_0^s}$ is a canonical symplectic form: $\omega|_{\Lambda_0^s} = dI \wedge d\varphi$.

Since $\Lambda_\varepsilon^s$ is $\varepsilon$-close to $\Lambda_0^s$ in the $\mathcal{C}^1$ sense, it follows that $\omega_\varepsilon^s \equiv \omega|_{\Lambda_\varepsilon^s}$ is non-degenerate and, since it is closed, it is a symplectic form.

Since the restriction of vector fields, forms and the exterior differential to a manifold behave in a functorial way, we have that the restriction of the Hamiltonian $\tilde{H}_\varepsilon$ to the manifold $\Lambda_\varepsilon^s \times \{s\}$ generates the vector field $\mathcal{H}_\varepsilon|_{\Lambda_\varepsilon^s \times \{s\}}$, where $\mathcal{H}_\varepsilon$ is the vector field of (3.4), restricted to the manifold using the symplectic form given by $\tilde{H}_\varepsilon$.

The map $\tilde{\mathcal{F}} = (\mathcal{F}, \mathrm{Id}_\mathbb{T})$ given by Theorem 7.1 allows us to define a system of coordinates in $\tilde{\Lambda}_\varepsilon$ by transporting the coordinates defined in $\tilde{\Lambda}$. Note that $\tilde{\mathcal{F}}$ satisfies normalization (7.3), so the coordinates $(I, \varphi, s)$ remain unaltered. We will use this to introduce symplectically conjugate coordinates as follows:

We can use $\mathcal{F}$ to pull back to $\Lambda$ the form $\omega_\varepsilon^s$. We note that $\mathcal{F}^* \omega_\varepsilon^s$ is close to $dI \wedge d\varphi$. Indeed, since the manifold $\Lambda_\varepsilon^s$ is 2-dimensional, we know that

$$(8.1) \qquad \qquad \mathcal{F}^* \omega_\varepsilon^s = a(I, \varphi, s; \varepsilon) dI \wedge d\varphi,$$

where

$$a(I, \varphi, s; \varepsilon) = 1 + \{\Pi_p \tilde{\mathcal{F}}, \Pi_q \tilde{\mathcal{F}}\} = 1 + \varepsilon^2 \{\Pi_p \tilde{\mathcal{F}}_1, \Pi_q \tilde{\mathcal{F}}_1\} + \cdots$$

is of class $C^{r-2}$. We claim in ii) that there is $J = \mathcal{J}(I, \varphi, s; \varepsilon)$, such that

$$(8.2) \qquad \mathcal{F}^* \omega_\varepsilon^s = dJ \wedge d\varphi.$$

Indeed, the only function $\mathcal{J}$ satisfying this requirement is, up to an additive function of $\varphi$,

$$(8.3) \qquad \mathcal{J}(I, \varphi, s; \varepsilon) = \int_0^I a(L, \varphi, s; \varepsilon) \, dL,$$

which is $C^{r-2}$. This additive function is identically zero to satisfy the normalization iii).

$\mathcal{J}$ is $C^1$-close to $I$ on compact sets, so we have that $(J, \varphi, s)$ will be a good system of coordinates on $\tilde{\Lambda}$.

We can now push forward this function $\mathcal{J}$ and use on $\tilde{\Lambda}_\varepsilon$ the coordinate system $(J, \varphi, s)$.

By Proposition 7.5, we know that $\tilde{\mathcal{F}}_1$ is a trigonometric polynomial, $\mathcal{N}(\tilde{\mathcal{F}}_1) = \mathcal{N}(h_1)$, and that $\tilde{\mathcal{F}}_2$ is a trigonometric polynomial and $\mathcal{N}(\tilde{\mathcal{F}}_2) \subset (N(h_1) + \mathcal{N}(h_1)) \cup \mathcal{N}(h_2)$. Since $\tilde{\mathcal{F}}_i$ are trigonometric polynomials, we obtain that $J = I + \varepsilon^2 \mathcal{J}_2(I, \varphi, s) + \cdots + \varepsilon^m \mathcal{J}_m(I, \varphi, s; \varepsilon)$, where $\mathcal{J}_i$ are trigonometric polynomials in $(\varphi, s)$, $\mathcal{N}(\mathcal{J}_2) \subset \mathcal{N}(\tilde{\mathcal{F}}_1) + \mathcal{N}(\tilde{\mathcal{F}}_1) \subset \mathcal{N}(h_1) + \mathcal{N}(h_1)$. $\qquad \square$

REMARK 8.3. Notice that the explicit formula (8.3) requires that $I$, $\varphi$ are one-dimensional.

Nevertheless, we point out that a similar result—with a higher loss of derivatives—can be established in higher dimensions using a global version of the Darboux Theorem, subject to the constraint that the change of coordinates does not change the variable $\varphi$.

Since the one-dimensional case is enough for the purposes of this paper, we will not purse the matter here.

## 8.2. Calculation of the reduced Hamiltonian

The goal of this section is to examine the reduced vector field $\mathcal{H}_\varepsilon|_{\Lambda_\varepsilon^s \times \{s\}}$ when written in the variables $(J, \varphi, s)$ given by Proposition 8.2.

PROPOSITION 8.4. *Let $k(J, \varphi, s; \varepsilon)$ be the Hamiltonian of the vector field $\mathcal{H}_\varepsilon$ restricted to $\tilde{\Lambda}_\varepsilon$, and expressed in the variables given by Proposition 8.2.*

*If $h$ is $C^r$ in all the variables and $r \geq 3$, then $k$ is $C^{r-2}$ in all the variables. Therefore, for $r - 2 > m \geq 0$, we can expand*

$$(8.4)$$
$$k(J, \varphi, s; \varepsilon) = k_0(J) + \varepsilon k_1(J, \varphi, s) + \cdots + \varepsilon^m k_m(J, \varphi, s) + \mathrm{O}_{C^{r-m-3}}(\varepsilon^{m+1})$$

*where $k_0(J) = \frac{J^2}{2}$, and $k_i$ are of class $C^{r-2-i}$.*

*In case that*

$$h(p, q, I, \varphi, s; \varepsilon) = h_1(p, q, I, \varphi, s) + \cdots + \varepsilon^{m-1} h_m(p, q, I, \varphi, s) + \mathrm{O}_{C^{r-m}}(\varepsilon^m),$$

*and $h_1, \ldots, h_m$ are trigonometric polynomials in $(\varphi, s)$, then*

i) *$k_1, \ldots, k_m$ are trigonometric polynomials in $(\varphi, s)$.*

ii)

$$(k_1, k_2) = \mathcal{G}(h_1|_{p=0,q=0}, h_2|_{p=0,q=0}, \nabla_{p,q} h_1|_{p=0,q=0}, \nabla_{p,q}^2 h_1|_{p=0,q=0})$$

*where $\mathcal{G}$ is a polynomial function which can be written explicitly.*

iii)

$$\mathcal{G}(0,0,0,0) = 0, \quad D_{h_1}\mathcal{G}(0,0,0,0) = D_{h_2}\mathcal{G}(0,0,0,0) = Id,$$

*and all the other partial derivatives of $\mathcal{G}$ evaluated at $(0,0,0,0)$ are zero.*

iv) $\mathcal{N}(k_1) \subset \mathcal{N}(h_1)$, *and* $\mathcal{N}(k_2) \subset \mathcal{N}(h_1) + \mathcal{N}(h_1) \cup \mathcal{N}(h_2)$.

Among the results of Proposition 8.4, the most important for us is that the restricted Hamiltonian is still a trigonometric polynomial up to order $\varepsilon^m$. The subsequent analysis will take some advantage of the fact that if we truncate to a fixed order in $\varepsilon$, we only need to deal with a Hamiltonian with a finite number of harmonics.

The analysis we will present later will have as hypothesis that $k_1, k_2$ avoid some manifolds of positive codimension in the space of trigonometric polynomials. The conclusion ii) of Proposition 8.4 shows that this hypothesis is implied by $h_1, h_2$ avoiding a manifold of positive codimension in the set of trigonometric polynomials. In particular, we conclude that the results of subsequent analysis are verified for a generic set of $h$'s.

PROOF. It is almost obvious using that, by Proposition 7.5, one can write rather explicitly a formal power series in $\varepsilon$ for the function $\tilde{\mathcal{F}}$, and that all the terms in the expansion are trigonometric polynomials.

We start by noting that since the Hamiltonian restricted to the invariant manifold generates the restricted flow when we consider the restricted form, we have that the function

$$K : [I_-, I_+] \times \mathbb{T} \times \mathbb{T} \times [-\varepsilon^*, \varepsilon^*] \to \mathbb{R}$$

defined by: $K(I, \varphi, s; \varepsilon) = H_\varepsilon \circ \tilde{\mathcal{F}}(I, \varphi, s; \varepsilon)$ defines the Hamiltonian as a function of the variables $(I, \varphi, s; \varepsilon)$.

Then, the desired Hamiltonian in the coordinate system $(J, \varphi, s)$ will be:

$$(8.5) \qquad k(J, \varphi, s; \varepsilon) = H_\varepsilon \circ \tilde{\mathcal{F}}(\mathcal{I}(J, \varphi, s; \varepsilon), \varphi, s; \varepsilon),$$

where we have denoted by $\mathcal{I}(J, \varphi, s; \varepsilon)$ the inverse of the function $I \mapsto \mathcal{J}(I, \varphi, s; \varepsilon)$ given in (8.3), which is a $\mathcal{C}^{r-2}$ function. Since the function $\mathcal{J}$ has an expansion in $\varepsilon$ given in Proposition 8.2 the function $\mathcal{I}$ has an expansion with similar properties.

Therefore, if we expand in $\varepsilon$ the expression for $k(J, \varphi, s; \varepsilon)$ given in (8.5), we obtain the results claimed in Proposition 8.4. Moreover,

$$(8.6) \qquad k_1(J, \varphi, s) = DH_0 \circ \tilde{\mathcal{F}}_0(J, \varphi, s) \tilde{\mathcal{F}}_1(J, \varphi, s) + h_1 \circ \tilde{\mathcal{F}}_0(J, \varphi, s),$$

and one can obtain explicit formulas for $k_2$, etc.

$\square$

## 8.3. Isolating the resonances (resonant averaging)

The main result of this section will be Theorem 8.9. This Theorem makes precise the notion that, when the perturbation is a trigonometric polynomial, we can find a finite set of resonances where the averaging breaks down. When we are at a distance $L$—which we will take as a fixed number—from the resonances we can perform averaging transformations and reduce the system very close to strongly integrable. Near the resonances, the averaging transformations reduce the system very close to an integrable pendulum.

The fact that we take $L$ as a fixed number is just to avoid having to carry out different estimates for different resonances. It is clearly not optimal except for the readability. The fact that we take $L$ independent of the resonance is the only reason why we take the perturbation to be a polynomial.

REMARK 8.5. Even if we hope that the treatment of this section is quite self-contained, we mention that very pedagogical treatments of averaging theory can be found in [**AKN88, LM88**]. Arguments very similar to those used here appear in the proof of KAM theorem using the strategy of [**Arn63a**]. See [**Sva80**] and specially [**DG96**]. The main difference between our study and that of KAM theory is that we will pay attention to the description of the phenomena that happen in the resonant regions. In KAM theory, the resonant regions are not analyzed and the only thing done with them is to estimate their size.

REMARK 8.6. We will perform the averaging procedure to a rather high order (i.e. $\varepsilon^{26}$) and to exclude from the resonances a size $L$ which is independent of the resonances and of the parameter $\varepsilon$.

Choosing the order in $\varepsilon$ to be as high is done to simplify the exposition later. As we will see the only resonances that produce effects that affect subsequent arguments are those of order 1 and 2 (we will show that resonances of order $j$ produce gaps among KAM tori of size $\varepsilon^{j/2}$ and there is a standard method to transverse those gaps of order smaller than $\varepsilon$). Hence, any order of averaging greater or equal than 2 would have been enough. Nevertheless, working harder at the averaging step will make other arguments in Section 8.5.2 simpler. We decided that this future simplification was worth that the very slight complication in the present chapter.

REMARK 8.7. Choosing the width of the resonant regions to be a fixed number $L$ is rather wasteful. It is well known that one can choose the width to be a suitable power of $\varepsilon$ multiplied by the size of the Fourier coefficients, which decrease with $|k|$.

For the purposes of this paper, choosing that width to be a constant is enough and simplifies the exposition. A more precise choice would be needed to eliminate the hypothesis **H3** from Theorem 4.1 that requires the perturbation to be a trigonometric polynomial. A heuristic discussion of these choices is included in Section 12.4.

**8.3.1. The infinitesimal equations for averaging.** Now we turn to the implementation of the above strategy. We will use the formalism of Lie series. Tutorials on this method appear in [**Car81, Mey91, Lla01**]. A comparison of different versions appears in [**LMM86**]. We will be considering canonical transformations obtained as the time-one map of a Hamiltonian.

We first start discussing the first order equations, which will serve as motivation for the phenomena of resonances.

We recall that if $g$ is the canonical transformation obtained by the time-one flow of a periodic in time Hamiltonian $\varepsilon G$ then, given a smooth function $K$, we have that $K \circ g = K + \varepsilon\{K, G\} + \mathrm{O}(\varepsilon^2)$ where $\{\ ,\ \}$ denotes the Poisson bracket in the canonical coordinates $(J, \varphi, A, s)$. If $K = K_0 + \varepsilon K_1$, we see that $K \circ g = K_0 + \varepsilon(K_1 + \{K_0, G\}) + \mathrm{O}(\varepsilon^2)$. Hence, when considering the transformations that make a perturbation simpler, it is natural to consider equations of the form:

$$(8.7) \qquad\qquad K + \{K_0, G\} = \bar{K},$$

where $K(J, \varphi, s)$ is a given Hamiltonian, $K_0 = A + \frac{J^2}{2}$, and the unknowns are $G$, the generator of the infinitesimal transformation, and $\bar{K}$, the averaged Hamiltonian, which will be chosen to be as simple as possible. Notice that, in this case, $\{K_0, G\} = J\frac{\partial G}{\partial \varphi} + \frac{\partial G}{\partial s}$ and then, if we write (8.7) in Fourier coefficients, we have,

$$K_{k,l}(J) + i(Jk + l)G_{k,l}(J) = \bar{K}_{k,l}(J),$$

where $K_{k,l}(J)$, $G_{k,l}(J)$, $\bar{K}_{k,l}(J)$, are the Fourier coefficients of $K$, $G$, $\bar{K}$, for $(k, l) \in \mathcal{N}$, the support of the Fourier transform of $K$.

When

$$(8.8) \qquad\qquad Jk + l = 0$$

we can not choose $G_{k,l}(J)$ in order to have $\bar{K}_{k,l}(J) = 0$. When $k = 0$, $l = 0$, equation (8.8) happens for all values of $J$, when $k = 0$, $l \neq 0$, it happens for no value of $J$. For all the cases $k \neq 0$, (8.8) happens when $J = -l/k$.

We will refer to resonances as the places $J = -l/k$, $(k, l) \in \mathcal{N}$, $k \neq 0$.

Notice that when $J$ is not resonant, we can reduce the system to contain only $\bar{K}_{0,0}(J)$, which is an integrable system.

To preserve smoothness, we will have also to keep a good part of these terms for neighboring values of the action. The tapering off of the elimination involves somewhat arbitrary choices.

In the following Lemma 8.8 we introduce a specific choice and provide estimates for it.

LEMMA 8.8. *Let*

$$K(J, \varphi, s) = \sum_{(k,l) \in \mathcal{N}} K_{k,l}(J) e^{i(k\varphi + ls)}$$

*be a Hamiltonian, with* $\mathcal{N} = \mathcal{N}(K) \subset \mathbb{Z}^2$ *a finite set. Assume that* $K$ *is of class* $\mathcal{C}^{n+1}$ *with respect to* $J$.

*Choose* $L$ *to be a number such that the real intervals*

$$[-l/k - 2L, -l/k + 2L] \quad (k,l) \in \mathcal{N}, k \neq 0$$

*are disjoint.*

*Then, there exist* $G$ *of class* $\mathcal{C}^n$ *with respect to* $J$, *and* $\bar{K}$ *of class* $\mathcal{C}^{n+1}$, *such that they solve the homological equation (8.7) and that verify:*

(1) *If* $|J + l/k| \geq 2L$ *for any* $(k,l) \in \mathcal{N}$, $k \neq 0$, *then*

$$\bar{K}(J, \varphi, s) = K_{0,0}(J).$$

(2) *If* $|J + l_0/k_0| \leq L$ *for some* $(k_0, l_0) \in \mathcal{N}$, $k_0 \neq 0$, *then*

$$\begin{aligned}
\bar{K}(J, \varphi, s) &= K_{0,0}(J) + \sum_{t=-N}^{N} K_{tk_0, tl_0}(-l_0/k_0) e^{it(k_0\varphi + l_0 s)} \\
&=: K_{0,0}(J) + U_{k_0, l_0}(k_0\varphi + l_0 s),
\end{aligned}$$

*where* $0 < N < \infty$ *is such that* $(tk_0, tl_0) \in \mathcal{N}$, *for* $|t| \leq N$.

(3) *The function* $\bar{K}$ *verifies:* $|\bar{K}|_{\mathcal{C}^{n+1}} \leq (1 + \frac{C}{L^{n+1}}) |K|_{\mathcal{C}^{n+1}}$, *where* $C$ *is a constant independent of* $L$.

(4) *The function* $G$ *verifies* $|G|_{\mathcal{C}^n} \leq \dfrac{C}{L^{n+1}} |K|_{\mathcal{C}^{n+1}}$.

(5) $\mathcal{N}(G)$ *and* $\mathcal{N}(\bar{K})$ *are finite sets. Moreover* $\mathcal{N}(G)$, $\mathcal{N}(\bar{K}) \subset \mathcal{N}(K)$.

PROOF. If we write (8.7) in Fourier coefficients, we have, for $(k,j) \in \mathcal{N}$:

(8.9) $$K_{k,l}(J) + i(Jk + l)G_{k,l}(J) = \bar{K}_{k,l}(J),$$

Then, in order to solve equation (8.9), we choose:

(1) If $(0,0) \in \mathcal{N}$, $\bar{K}_{0,0}(J) = K_{0,0}(J)$.
(2) If $(0,l) \in \mathcal{N}$, $l \neq 0$, $\bar{K}_{0,l}(J) = 0$.
(3) If $(0,0) \neq (k,l) \in \mathcal{N}$, $k \neq 0$, we choose $\bar{K}_{k,l}(J)$ as:

$$\bar{K}_{k,l}(J) = K_{k,l}(-l/k)\psi\left(\frac{1}{L}(J + l/k)\right)$$

where $\psi(t)$ is a fixed $\mathcal{C}^\infty$ function such that: $\psi(t) = 1$, if $t \in [-1, 1]$, and $\psi(t) = 0$, if $t \notin [-2, 2]$. With this choice we have that $\bar{K}_{k,l}$ verifies:

(a) If $|J + l/k| \leq L$ then $\bar{K}_{k,l}(J) = K_{k,l}(-l/k)$.
(b) if $|J + l/k| \geq 2L$ then $\bar{K}_{k,l}(J) = 0$.

Once we have defined $\bar{K}$ as above, it is clear that it is a $\mathcal{C}^{n+1}$ function, and that it verifies the desired bounds, where the constant $C$ only depends on the cut-off function $\psi$, and on $n$.

Now, we choose $G$ to verify equation (8.9):

(1) $G_{0,0}(J) = 0$, and $G_{k,l}(J) = 0$ if $(k, l) \notin \mathcal{N}$.

(2) $G_{0,l}(J) = -\dfrac{K_{0,l}(J)}{il}$,

(3) If $(k, l) \in \mathcal{N}$, $k \neq 0$, we choose $G_{k,l}(J)$ as:

    (a) If $J \neq -\dfrac{l}{k}$ then $G_{k,l}(J) = \dfrac{\bar{K}_{k,l}(J) - K_{k,l}(J)}{i(Jk + l)}$,

    (b) $G_{k,l}(-l/k) = \displaystyle\lim_{J \to -l/k} \dfrac{\bar{K}_{k,l}(J) - K_{k,l}(J)}{i(Jk + l)} = \dfrac{-K'_{k,l}(-l/k)}{ik}$.

To bound the function $G$, we note that, given a fixed $(k_0, l_0) \in \mathcal{N}$, by the definition of $\bar{K}$ and $G$, we have:

(1) $\forall J$, then $|G_{0,l}(J)|_{\mathcal{C}^n} \leq C \dfrac{|K_{0,l}|_{\mathcal{C}^n}}{|l|}$.

(2) If $|J + l_0/k_0| \leq L$ then $|G_{k_0,l_0}(J)|_{\mathcal{C}^n} \leq C \dfrac{|K_{k_0,l_0}|_{\mathcal{C}^{n+1}}}{|k_0|}$.

(3) If $|J + l_0/k_0| \geq 2L$ then $|G_{k_0,l_0}(J)|_{\mathcal{C}^n} \leq C \dfrac{|K_{k_0,l_0}|_{\mathcal{C}^n}}{|k_0| (2L)^{n+1}}$.

(4) If $L \leq |J + l_0/k_0| \leq 2L$ then

$$|G_{k_0,l_0}(J)|_{\mathcal{C}^n} \leq C \dfrac{|K_{k_0,l_0}(-l_0/k_0)| \, |\psi|_{\mathcal{C}^n} + |K_{k_0,l_0}|_{\mathcal{C}^n}}{|k_0| \, L^n}.$$

Then, $G(J, \varphi, s)$ is a trigonometric polynomial in $(\varphi, s)$, and of class $\mathcal{C}^n$ with respect to $J$. $\qquad\qquad\qquad\qquad\qquad\qquad\qquad\qquad\qquad\qquad\qquad\qquad\qquad \square$

**8.3.2. The main averaging result, Theorem 8.9.** Once we know how to solve any homological equation (8.7), we can proceed to obtain a suitable global normal form of our reduced Hamiltonian by applying repeatedly the procedure. The precise result is formulated in the following Theorem 8.9.

THEOREM 8.9. *Let $k(J, \varphi, s; \varepsilon)$ be a $\mathcal{C}^n$ Hamiltonian, $n > 1$, and consider any $1 \leq m < n$, independent of $\varepsilon$. Assume that*

(8.10) $$k(J, \varphi, s; \varepsilon) = \frac{J^2}{2} + \varepsilon k^1(J, \varphi, s; \varepsilon).$$

*Let $k_i(J, \varphi, s)$ $i = 1, \ldots, m$ be the coefficients in the Taylor expansion with respect to $\varepsilon$ of $k^1(J, \varphi, s; \varepsilon)$, and assume that the $k_i(J, \varphi, s)$, $i = 1, \ldots, m$ are trigonometric polynomials in $\varphi, s$.*

*Then, there exists a finite set (of resonances)*

$$\mathcal{R} = \mathcal{R}_1 \cup \cdots \cup \mathcal{R}_m \subset \mathbb{Q}$$

*(we will give rather explicit expressions for $\mathcal{R}_i$ involving the support of the Fourier transform of the $k_i$) such that:*

*Given $L$ a number, independent of $\varepsilon$, such that the real intervals $[-l/k - 2L, l/k + 2L]$ for $-l/k \in \mathcal{R}$, are disjoint, there exists a symplectic change of variables, depending on time, $(\mathcal{B}, \alpha, s) \mapsto (J, \varphi, s)$, periodic in $\varphi$ and $s$, and of class $\mathcal{C}^{n-2m}$, which is $\varepsilon$-close to the identity in the $\mathcal{C}^{n-2m-1}$ sense,* *such that transforms the Hamiltonian system associated to $k(J, \varphi, s; \varepsilon)$ into a Hamiltonian system of Hamiltonian*

$$\bar{k}(\mathcal{B}, \alpha, s; \varepsilon) = \bar{k}^0(\mathcal{B}, \alpha, s; \varepsilon) + \varepsilon^{m+1}\bar{k}^1(\mathcal{B}, \alpha, s; \varepsilon)$$

*where the function $\bar{k}^0$ is of class $\mathcal{C}^{n-2m+2}$, and $\varepsilon^{m+1}\bar{k}^1$ is of class $\mathcal{C}^{n-2m}$ and they verify:*

(1) *If $|\mathcal{B} + l/k| \geq 2L$ for any $(k, l) \in \mathbb{Z}^2$ such that $-l/k \in \mathcal{R}$, then*

$$\bar{k}^0(\mathcal{B}, \alpha, s; \varepsilon) = \frac{1}{2}\mathcal{B}^2 + \varepsilon\bar{k}^{0,0}(\mathcal{B}; \varepsilon)$$

*where $\bar{k}^{0,0}(\mathcal{B}; \varepsilon)$ is a polynomial of degree $m-1$ in $\varepsilon$.*

(2) *If $|\mathcal{B} + l_1/k_1| \leq L$ for some $(k_1, l_1) \in \mathbb{Z}^2$, and for some $1 \leq i \leq m$ we have that $-l_1/k_1 \in \mathcal{R}_i \setminus (\mathcal{R}_1 \cup \cdots \cup \mathcal{R}_{i-1})$, then*

$$\bar{k}^0(\mathcal{B}, \alpha, s; \varepsilon) = \frac{1}{2}\mathcal{B}^2 + \varepsilon\bar{k}^{0,0}(\mathcal{B}; \varepsilon) + \varepsilon^i U^{k_1, l_1}(k_1\alpha + l_1 s; \varepsilon)$$

*where $U^{k_1, l_1}(\theta; \varepsilon)$ is a polynomial in $\varepsilon$ and a trigonometric polynomial in $\theta$.*

(3) *If $|\mathcal{B} + l_0/k_0| \leq L$ for some $(k_0, l_0) \in \mathbb{Z}^2$ and $-l_0/k_0 \in \mathcal{R}_1$, then the function $U^{k_0, l_0}$ defined in 2 is given by:*

(8.11)
$$U^{k_0, l_0}(\theta; \varepsilon) = \sum_{t=-N}^{N} \hat{k}^1_{tk_0, tl_0}(-l_0/k_0; 0)e^{it\theta}$$
$$+ \mathrm{O}(\varepsilon)$$

*where $\hat{k}^1_{k,l}(J; \varepsilon)$ are the Fourier coefficients of the $k^1(J, \varphi, s; \varepsilon)$ with respect to the angle variables $(\varphi, s)$.*

Note that in this Theorem we have not claimed anything in the regions $L < |J + k/l| < 2L$. This is not a problem because, by remembering that $L$ is arbitrary, we can obtain the same results using $L/2$ in place of $L$. Hence, the analysis that we will carry out in each of the different pieces applies to the whole space. See the Remark 8.45 for more details.

**8.3.3. Proof of Theorem 8.9.** Theorem 8.9 follows from a repeated application of the following inductive Lemma 8.10, that we will prove using the method of Lie transforms [**Car81, Mey91, Lla01**].

The hypothesis of Lemma 8.10 are that we have a Hamiltonian already in normal form outside of a set or resonances. The conclusions are that, excluding an slightly larger set of resonances—which we will give rather explicitly—we can produce another Hamiltonian which is normalized to a higher order in $\varepsilon$.

LEMMA 8.10. *Consider a Hamiltonian of the form:*

(8.12)
$$k_q(J, \varphi, s; \varepsilon) = k_q^0(J, \varphi, s; \varepsilon) + \varepsilon^{q+1}k_q^1(J, \varphi, s; \varepsilon)$$

*where*

1. $k_0^0(J, \varphi, s; \varepsilon) = \frac{J^2}{2}$ and, if $q \geq 1$, $k_q^0(J, \varphi, s; \varepsilon)$ is a $\mathcal{C}^{n+2-2q}$ function that verifies:

   There exist finite sets $\mathcal{R}_i \subset \mathbb{Q}$, called resonances of order $i$, $i = 1 \cdots q$, and a number $L > 0$ such that:

   1.0. The intervals $\mathcal{I}_{-l/k} \equiv [-l/k - 2L, -l/k + 2L]$, $-l/k \in \mathcal{R}^{[\leq q]} \equiv \bigcup_{i=1,\ldots,q} \mathcal{R}_i$ are disjoint.

   1.1. If $J \notin \bigcup_{-l/k \in \mathcal{R}^{[\leq q]}} \mathcal{I}_{-l/k}$, then

   $$k_q^0(J, \varphi, s; \varepsilon) = \frac{J^2}{2} + \varepsilon k_q^{0,0}(J; \varepsilon),$$

   where $\varepsilon k_q^{0,0}(J; \varepsilon)$ is a polynomial of degree $q$ in $\varepsilon$.

   1.2. If $|J + l_1/k_1| \leq L$ for some $(k_1, l_1) \in \mathbb{Z}^2$ such that

   $$-l_1/k_1 \in \mathcal{R}_i \setminus (\mathcal{R}_1 \cup \cdots \cup \mathcal{R}_{i-1}),$$

   for some $1 \leq i \leq q$, then

   $$k_q^0(J, \varphi, s; \varepsilon) = \frac{J^2}{2} + \varepsilon k_q^{0,0}(J; \varepsilon) + \varepsilon^i U_q^{k_1, l_1}(k_1 \varphi + l_1 s; \varepsilon)$$

   where $U_q^{k_1, l_1}(\theta; \varepsilon)$ is a polynomial in $\varepsilon$ and a trigonometric polynomial in $\theta \equiv k_1 \varphi + l_1 s$.

2. $k_q^1(J, \varphi, s; \varepsilon)$ is a $\mathcal{C}^{n-2q}$ function whose Taylor series coefficients with respect to $\varepsilon$ are trigonometric polynomials in $(\varphi, s)$.

Denote by $K = k_q^1(J, \varphi, s; 0)$, which is the term of the perturbation of order exactly $q + 1$. Introduce also the set

(8.13)         $\mathcal{R}_{q+1} = \{-l/k, \ (k, l) \in \mathcal{N}(k_q^1(\cdot; 0)), \ k \neq 0\}.$

Choose $0 < \tilde{L} < L$ such that the intervals $[-l/k - 2\tilde{L}, -l/k + 2\tilde{L}]$ are disjoint when $-l/k \in \mathcal{R}^{[\leq q+1]}$.

Let $G(\mathcal{B}, \alpha, s)$ be the $\mathcal{C}^{n-2q-1}$ function given by Lemma 8.8, verifying equation (8.7) with $K = k_q^1(J, \varphi, s; 0)$ and with a distance $\tilde{L}$ away from the resonances.

Then, the $\mathcal{C}^{n-2q-2}$ change of variables

$$(J, \varphi, s) = g(\mathcal{B}, \alpha, s),$$

given by the time one flow of the Hamiltonian $\varepsilon^{q+1} G(\mathcal{B}, \alpha, s)$, transforms the Hamiltonian $k_q(J, \varphi, s; \varepsilon)$ into a Hamiltonian

$$k_{q+1}(\mathcal{B}, \alpha, s; \varepsilon) = k_{q+1}^0(\mathcal{B}, \alpha, s; \varepsilon) + \varepsilon^{q+2} k_{q+1}^1(\mathcal{B}, \alpha, s; \varepsilon),$$

with

(8.14)         $k_{q+1}^0(\mathcal{B}, \alpha, s; \varepsilon) = k_q^0(\mathcal{B}, \alpha, s; \varepsilon) + \varepsilon^{q+1} \bar{k}_q^1(\mathcal{B}, \alpha, s; 0),$

where $\bar{k}_q^1(\mathcal{B}, \alpha, s; 0) = \bar{K}(\mathcal{B}, \alpha, s)$, given in Lemma 8.8, is a $\mathcal{C}^{n-2q}$ function.

Moreover, the Hamiltonian $k_{q+1}^0(\mathcal{B}, \alpha, s; \varepsilon)$ verifies properties [1.0], [1.1], [1.2] up to order $q + 1$ with $\tilde{L}$ replacing $L$.

*Furthermore, $\varepsilon^{q+2}k_{q+1}^1(\mathcal{B}, \alpha, s; \varepsilon)$ is a $\mathcal{C}^{n-2q-2}$ function whose Taylor series coefficients with respect to $\varepsilon$ are trigonometric polynomials in $(\varphi, s)$.*

PROOF. We use Lemma 8.8 to solve equation (8.7), for the function $K = k_q^1(J, \varphi, s; 0)$, which is a trigonometric polynomial in $(\varphi, s)$. We obtain a $\mathcal{C}^{n-2q-1}$ function $G(\mathcal{B}, \alpha, s)$ and a $\mathcal{C}^{n-2q}$ function $\bar{k}_q^1(\mathcal{B}, \alpha, s; 0)$. These solutions are also trigonometric polynomials in $(\alpha, s)$, and verify the assumptions of Lemma 8.8 with the finite resonance set $\mathcal{R}_{q+1}$ defined in (8.13).

After applying the $\mathcal{C}^{n-2q-2}$ transformation $g$ defined as the time one flow of the Hamiltonian $\varepsilon^{q+1}G$ to the extended autonomous Hamiltonian $A + k_q$, the new Hamiltonian is given by

$$
\begin{aligned}
A + k_{q+1} &= (A + k_q) \circ g \\
&= (A + k_q^0) \circ g + \varepsilon^{q+1}k_q^1 \circ g \\
&= A + k_q^0 + \varepsilon^{q+1}\left(\{k_0^0 + A, G\} + k_q^1(\cdot; 0)\right) \\
&\quad + \varepsilon^{q+2}k_{q+1}^1,
\end{aligned}
$$

where $k_0^0 = \frac{J^2}{2}$. Using the homological equation verified by $G$, we get that the Hamiltonian $k_{q+1}^0$ in (8.14) verifies properties [1.0], [1.1], [1.2], up to order $q + 1$, and using that $k_q^0 = k_0^0 + O_{\mathcal{C}^{n+1-2q}}(\varepsilon)$, we obtain that

$$
\begin{aligned}
\varepsilon^{q+2}k_{q+1}^1 &= (A + k_0^0) \circ g - (A + k_0^0) - \{(A + k_0^0), \varepsilon^{q+1}G\} \\
&\quad + (k_q^0 - k_0^0) \circ g - (k_q^0 - k_0^0) \\
&\quad + \varepsilon^{q+1}\left(k_q^1 \circ g - k_q^1\right) + \varepsilon^{q+1}[k_q^1 - k_q^1(\cdot; 0)]
\end{aligned}
$$

is a bounded $\mathcal{C}^{n-2q-2}$ function because $k_q^0$ is a $\mathcal{C}^{n-2q+2}$ function, $k_q^1$ is a $\mathcal{C}^{n-2q}$ function, and $G$ is analytic with respect to $(\varphi, s)$.

Moreover, all the terms in the Taylor series of $k_{q+1}^1(\mathcal{B}, \alpha, s; \varepsilon)$ with respect to $\varepsilon$, are obtained from a finite number of algebraic operations and Poisson brackets between the Taylor coefficients in $\varepsilon$ of $k_q^0$, those of $k_q^1$ and those of $G$.

Since all the Taylor coefficients of $k_q^1$ are trigonometric polynomials in the angle variables, the $G$ provided by Lemma 8.8 is a trigonometric polynomial in the angular variables.

Hence, we conclude that the Taylor coefficients of $k_{q+1}^1(\mathcal{B}, \alpha, s; \varepsilon)$ with respect to $\varepsilon$ are also trigonometric polynomials in the angle variables.    $\square$

PROOF OF THEOREM 8.9. The proof is by induction in $q$. To begin the induction we apply the inductive Lemma 8.10 for $q = 0$ to our Hamiltonian 8.10, with $k_0^0 = J^2/2$ and $\varepsilon k_0^1(J, \varphi, s; \varepsilon) = \varepsilon k^1(J, \varphi, s; \varepsilon)$, which is a $\mathcal{C}^n$ function. The set $\mathcal{R}_1 = \{J = -l/k, \ (k, l) \in \mathcal{N}(k^1(\cdot; 0))\}$, by hypothesis, is a finite set.

Then the change of variables $g$ provided by Lemma 8.10 is of class $\mathcal{C}^{n-2}$, and we obtain $k_1 = k_1^0 + \varepsilon^2 k_1^1$, where $k_1^0$ is a $\mathcal{C}^n$ function, and $\varepsilon^2 k_1^1$ is a $\mathcal{C}^{n-2}$ function, verifying properties [1.1], [1.2] of Lemma 8.10 with $q = 1$. Applying Lemma 8.8 one obtains formula (8.11).

Once we have the normal form up to order $q$, Lemma 8.10 gives us the normal form up to order $q + 1$ modifying the constant $L$ to accommodate new resonances. We only need to observe that, if $q \geq 1$, the modifications to the Hamiltonian $k_q^0$ are given in the terms of order $O(\varepsilon^{q+1})$, so the expression (8.11) remains invariant in all the process.

Applying the inductive Lemma 8.10 $m$ times, we obtain Theorem 8.9.

$\square$

REMARK 8.11. It is clear that we can apply Theorem 8.9 to our Hamiltonian $k(J, \varphi, s; \varepsilon)$ in (8.4) given in Proposition 8.4, with $n = r - 2$, and, after $m$ steps of averaging we will obtain a Hamiltonian $\bar{k}(\mathcal{B}, \alpha, s; \varepsilon) = \bar{k}_0(\mathcal{B}, \alpha, s; \varepsilon) + \varepsilon^{m+1}\bar{k}_1(\mathcal{B}, \alpha, s; \varepsilon)$, where $\bar{k}_0(\mathcal{B}, \alpha, s; \varepsilon)$ is of class $\mathcal{C}^{r-2m}$ and $\varepsilon^{m+1}\bar{k}_1(\mathcal{B}, \alpha, s; \varepsilon)$ is of class $\mathcal{C}^{r-2m-2}$.

## 8.4. The non-resonant region (KAM theorem)

In this section we analyze the Hamiltonian in the non-resonant region $\mathcal{S}^L$ of Theorem 8.9. That is, we study the region where the action $J$ is far from any resonance of order less or equal than $m$:

$$(8.15) \quad \mathcal{S}^L = \{(J, \varphi, s) \in \tilde{\Lambda}_\varepsilon : \; |J + l/k| \geq 2L, \; (k, l) \in \mathbb{Z}^2 \text{ for } -l/k \in \mathcal{R}\}$$

where $\mathcal{R} = \mathcal{R}_1 \cup \cdots \cup \mathcal{R}_m$, and $L$ are provided by Theorem 8.9.

By Theorem 8.9, in the averaged variables $(\mathcal{B}, \alpha, s)$ defined in the connected components of $\mathcal{S}^L$, we can write the Hamiltonian as

$$(8.16) \qquad \bar{k}(\mathcal{B}, \alpha, s; \varepsilon) = \frac{\mathcal{B}^2}{2} + \varepsilon\bar{k}^{0,0}(\mathcal{B}; \varepsilon) + \varepsilon^{m+1}\bar{k}^1(\mathcal{B}, \alpha, s; \varepsilon).$$

The point that will be most important for us is that the first two terms of (8.16) correspond to a strongly integrable Hamiltonian, which moreover, satisfies twist conditions. The other term is extremely small. Hence, using quantitative versions of KAM theorem, we will conclude, in the following quantitative Theorem, that the non-resonant region contains KAM tori which are $O(\varepsilon^{(m+1)/2)})$-closely spaced.

THEOREM 8.12. Let $\bar{k}$ be a Hamiltonian of the form (8.16), where $\varepsilon$ is sufficiently small and fixed. Assume that

    i) $\bar{k}$ is a $C^{s+\beta}$ function and $\bar{k}^{0,0}$ is a $C^{s+\beta+2}$ function of the variables $B, \alpha, s$, and that $\|k^{0,0}\|_{C^{s+\beta+2}}, \|k^1\|_{C^{s+\beta}}$ are bounded independently of $\varepsilon$.

    ii) We have $s \geq 5$, $0 < \beta < 1$.

Then we can find $C > 0$, depending only on the properties of the Hamiltonian but independent of $\varepsilon$ such that

    a) For every interval $\mathcal{I}$ of length $C\varepsilon^{(m+1)/2}$ there is an invariant torus $\mathcal{T}$ for the Hamiltonian flow contained in $\mathcal{I} \times \mathbb{T}^1 \times \mathbb{T}^1$.

b) *The torus $\mathcal{T}$ is the graph of a $C^{s-2+\beta}$ function $\Psi$ from the angle variables to the action-variables*

$$\mathcal{T} = \{(B, \alpha, s) \in \mathcal{I} \times \mathbb{T}^2; B = \Psi(\alpha, s)\}.$$

c) *We have, for some constant $a$,*

$$\|\Psi - a\|_{C^{s-2+\beta}} \leq C\varepsilon^{(m+1)/2}.$$

d) *The motion on the torus is $C^{s-4+\beta}$ conjugate to a rigid translation of frequency $(\omega, 1)$, where $\omega$ is a Diophantine number of constant type with Markov constant less that $\varepsilon^{(m+1)/2}$ (see Definition 8.16).*

REMARK 8.13. It is important to note that, for a fixed value of $\varepsilon$, as long as we fix $m > 2$, it is enough to consider a finite number of tori $\mathcal{T}_i$ to ensure that all the points in the non-resonant region $\mathcal{S}^L$ are $O(\varepsilon^{1+\delta})$-close, $\delta > 0$, to an invariant torus. Of course, this number goes to infinity when $\varepsilon$ decreases to zero.

The main technical tool to establish Theorem 8.12 is a quantitative version of KAM theorem (Theorem 8.19) which makes explicit the dependence of the Diophantine constants of the frequencies of tori with the size of the perturbations under which they can survive. We will also need to discuss the spacings between numbers with appropriate Diophantine properties in Section 8.4.1.

Theorem 8.19 is a general KAM theorem which will also be useful later in the proof of Theorem 8.30. Theorem 8.19 is just an adaptation to our notation of the theorem on [**Her83**, p. 198]. At the end of Section 8.4.2, we will show how Theorem 8.12 can be deduced from Theorem 8.19 and Lemma 8.17.

Note that, since we are using systems that are one-degree of freedom systems depending periodically on time, we can take time-$2\pi$ maps and reduce the problem to the study of a twist map of an annulus. It is easy to see that the invariant circles for the twist map correspond exactly to invariant tori for the Hamiltonian system.

The result of [**Her83**, p. 198], besides improving the differentiability requirements from other theorems, has the advantage that it contains very good results on the dependence on the Diophantine constants, which will lead to close spacing among the circles.

REMARK 8.14. The theorem in [**Her83**] has the disadvantages that it does not generalize to higher dimensions and it requires that the rotation numbers are of constant type (see Definition 8.16), hence it cannot establish that the tori cover a set of positive measure. For our purposes, however, measure does not play any role while geometric properties such as maximal gaps do.

REMARK 8.15. Since there are many versions of KAM theorem, differing in the differentiability required or smallness conditions, it is worth point out that optimizing these possibilities is not very crucial for us.

The smallness condition can always be adjusted by making more averaging transformations, which also amounts to require more differentiability.

**8.4.1. Some results on Diophantine approximation.** In this section we recall some results on Diophantine approximations, notably we discuss how separated are the numbers with good Diophantine constants. When incorporated to a KAM theorem this will lead to estimates on the gaps between the KAM tori. Later on, it will lead to estimates on how close to a resonance we can get tori.

We start by recalling the standard definition:

DEFINITION 8.16. *We say that a real number $\omega \in \mathcal{D}(\kappa, \nu)$ when we have:*

$$(8.17) \qquad |\omega - \ell/k| \geq \kappa \, |k|^{-2-\nu} \quad \forall (k, \ell) \in \mathbb{Z}^2, \; k \neq 0.$$

*We will refer to $\mathcal{D}(\kappa, \nu)$ as the Diophantine numbers of type $\kappa, \nu$. When $\nu = 0$, we will refer to $\mathcal{D}(\kappa, 0)$ as constant type numbers of Markov constant $\kappa$.*

*We will denote $\mathcal{D}(\nu) = \cup_{\kappa > 0} \mathcal{D}(\kappa, \nu)$ and refer to them as Diophantine numbers of type $\nu$. When $\nu = 0$ we will call them constant type numbers.*

The Definition 8.16 is the same as that used in [**Her83**, p. 158]. We note that the inequality (8.17) is equivalent to

$$(8.18) \qquad |\omega k - \ell|^{-1} \leq C\kappa^{-1} \, |(k, l)|^{1+\nu} \quad \forall (k, \ell) \in \mathbb{Z}^2$$

which is a form that generalizes to higher dimensions.

It is well known that $\mathcal{D}(\nu)$ is of full measure when $\nu > 0$. Constant type numbers are of zero measure but are dense. All quadratic irrationals are constant type numbers. More generally, it is equivalent to say that a number is constant type and to say that its continued fraction expansion is bounded. See Proposition 8.18 later.

The following statement makes it precise the idea that if $\kappa$ is small, the numbers $\mathcal{D}(\kappa, \nu)$, $\nu \geq 0$, do not have big gaps among them. The result that we will use later is the case for $\nu = 0$.

LEMMA 8.17. *Given $\nu \geq 0$, there exists a constant $K(\nu) > 0$ such that in every interval $\mathcal{I}$ of diameter $K(\nu)\kappa$ there is a number in $\mathcal{D}(\kappa, \nu)$.*

PROOF. There is a standard proof for $\nu > 0$ which also gives information on the measure and also generalizes for higher dimensions. We do it first.

We note that

$$\mathcal{D}(\kappa, \nu) = \mathbb{R} \setminus \bigcup_{(a,b) \in \mathbb{Z}^2, b > 0} B_{\kappa b^{-2-\nu}}(a/b)$$

Hence, given an interval $\mathcal{I}$ and denoting by $|\ \ |$ the Lebesgue measure of a set, we have

$$|\mathcal{I} \cap \mathcal{D}(\kappa, \nu)| \geq |\mathcal{I}| - 2 \sum_{(a,b) \in \mathbb{Z}^2, b>0, B_{\kappa|b|^{-2-\nu}}(a/b) \cap \mathcal{I} \neq \emptyset} \kappa |b|^{-2-\nu}$$

$$(8.19) \qquad \geq |\mathcal{I}| - 2 \sum_{b>0} \left( |\mathcal{I}| \kappa |b|^{-1-\nu} + 2\kappa |b|^{-2-\nu} \right)$$

$$\geq |\mathcal{I}| (1 - K_1 \kappa) - \kappa K_2.$$

The second inequality in (8.19) follows from the observation that, once we fix $b$, the number of $a$ such that $B_{\kappa b^{-2-\nu}}(a/b) \cap \mathcal{I} \neq \emptyset$ is bounded by $|\mathcal{I}| |b| + 2$. In the last inequality of (8.19), $K_1, K_2$ stand for positive constants that depend on $\nu$.

We see that for $\kappa$ small enough and for $|\mathcal{I}| \geq K\kappa$, the right hand side of (8.19) is positive, hence, the measure $\mathcal{I} \cap \mathcal{D}(\kappa, \nu)$ is positive, which establishes the claim.

The above derivation uses essentially that $\nu > 0$ since we use that $\sum_{b>0} b^{-1-\nu}$ converges. Indeed, it is well known that $\mathcal{D}(\kappa, 0)$ is of measure 0 for all $\kappa$.

For the case when $\nu = 0$, we will use the theory of continued fractions. We recall the following proposition whose proof can be found in [**Her79**, p. 64].

PROPOSITION 8.18. *Given $L \in \mathbb{R}^+$, denote by*

$$(8.20) \qquad \mathcal{C}(L) = \{x = [a_1, a_2, \ldots, a_n, \ldots] : \ a_i \in \mathbb{N}, \ 1 \leq a_i \leq L\}.$$

*With this notation, we have*

$$\mathcal{C}(L) = \mathcal{D}(1/(L+2), 0).$$

Hence, to finish the proof of Lemma 8.17, it suffices to show that $\mathcal{I}_L$, the largest interval in $\mathbb{R} \setminus \mathcal{C}(L)$, satisfies

$$(8.21) \qquad |\mathcal{I}_L| \leq K/L.$$

We claim that this interval $\mathcal{I}_L$, which is the largest gap among numbers in $\mathcal{C}(L)$, is precisely:

$$(8.22) \qquad \mathcal{I}_L = ([2, L, 1, L, 1, L, \ldots], [1, 1, L, 1, L, 1, \ldots])$$

Once we have the claim (8.22), the result in Lemma 8.17 follows because a direct calculation shows that

$$[2, L, 1, L, 1, \ldots] = 1/2 - 1/L + O(1/L^2)$$

and

$$[1, 1, L, 1, L, 1, \ldots] = 1/2 + 1/L + O(1/L^2).$$

Therefore $|\mathcal{I}_L| = 2/L + O(1/L^2)$. Hence, using Proposition 8.18, the largest gap in $\mathcal{D}(1/(L+2), 0)$ is $2/L + O(1/L^2)$. That is, we have established Lemma 8.17 for $\kappa = 1/(L+2)$, from which the general result follows.

Hence, the only thing left for the proof of Lemma 8.17 is to establish (8.22).

We recall that given two numbers

$$x = [a_1, a_2, \ldots, a_m, b_1, b_2, \ldots],$$
$$y = [a_1, a_2, \ldots, a_m, c_1, c_2, \ldots],$$

and $b_1 \neq c_1$, we have that $x > y$ if $b_1 > c_1$ and $m$ is odd or if $b_1 < c_1$ and $m$ is even.

The above observation allows to conclude immediately that the number $[2, L, 1, L, 1, L, \ldots]$ is the largest number in $\mathcal{C}(L)$ whose first entry in the continued fraction is 2. Similarly, the number $[1, 1, L, 1, L, 1, \ldots]$ is the smallest number in $\mathcal{C}(L)$ whose first entry is 1. This makes it clear that there are no points of $\mathcal{C}(L)$ inside $\mathcal{I}_L$.

The claim (8.22) follows from the following considerations.

(1) If the interval

$$\mathcal{I} = ([a_1, b_2, b_3, b_4, \ldots], [a_1, c_2, c_3, c_4, \ldots])$$

does not contain any point in $\mathcal{C}(L)$, then, neither does the interval

$$\hat{\mathcal{I}} = ([c_2, c_3, c_4, \ldots], [b_2, b_3, b_4, \ldots]).$$

Obviously, the interval $\hat{\mathcal{I}}$ is larger than the interval $\mathcal{I}$. Hence we conclude that the interval $\mathcal{I}_L$ has to have different first entries.

(2) If $M, N \in \mathbb{N}$ and $N > 1$, we have:

$$[M + 1, b_2, b_3, b_4, \ldots] \in ([M + N, b_2, b_3, b_4, \ldots], [M, c_2, c_3, c_4, \ldots]).$$

Hence, $\mathcal{I}_L$ should be of the form

$$\mathcal{I}_L = ([M + 1, b_2, b_3, b_4, \ldots], [M, c_2, c_3, c_4, \ldots]).$$

(3) The length of the interval $([2, b_2, b_3, b_4, \ldots], [1, c_2, c_3, c_4, \ldots])$ is larger than that of the interval $([M+1, b_2, b_3, b_4, \ldots], [M, c_2, c_3, c_4, \ldots])$ for $M > 1$.

Hence, $\mathcal{I}_L$ should be of the form

$$\mathcal{I}_L = ([2, b_2, b_3, b_4, \ldots], [1, c_2, c_3, c_4, \ldots]).$$

(4) If an interval of the form $([2, b_2, b_3, b_4, \ldots], [1, c_2, c_3, c_4, \ldots])$ does not contain any point in $\mathcal{C}(L)$, then, it is necessary that $b_2, b_3, \ldots$ are chosen in such a way that $[2, b_2, b_3, b_4, \ldots]$ is the largest number in $\mathcal{C}(L)$ whose first entry in the continued fraction expansion is 2. Similarly, we also need that $c_2, c_3, c_4, \ldots$ are chosen in such a way that $[1, c_2, c_3, c_4, \ldots]$ is the smallest number in $\mathcal{C}(L)$ whose first entry in the continued fraction expansion is 1.

Hence, we conclude that $\mathcal{I}_L$ is of the form given in (8.22).

$\square$

**8.4.2. The KAM Theorem for twist maps.** The following result is an easy consequence of the Theorem 5.4 stated in [**Her83**, p. 198] (see also the Theorem 5.6 in [**Her83**, p. 204]).

THEOREM 8.19. *Let $F_0 : \mathbb{T} \times \mathbb{R}$ be an integrable symplectic mapping, that is:*

$$(8.23) \qquad\qquad F_0(\theta, r) = (\theta + \Delta(r), r).$$

*Assume that $F_0 \in C^{n+\beta}$, $n \geq 3$, $0 < \beta < 1$ and $\frac{d}{dr}\Delta(r) \geq M > 0$.*

*Then we can find a constant $K$ depending only on $n, \beta$ such that for any $r_0$ such that $\omega \equiv \Delta(r_0) \in \mathcal{D}(\kappa, 0)$, and for any $F$ exact symplectic $C^{n+\beta}$ map of $\mathbb{T} \times \mathbb{R}$ verifying*

    i) $||F - F_0||_{C^{n+\beta}} \leq \delta$
    ii) $\delta \leq K\kappa(\omega)^2$,

*there exists an invariant circle $\mathcal{T}$ such that*

    a) $\mathcal{T}$ *is the graph of a function* $\Psi : \mathbb{T} \to \mathbb{R}$

$$\mathcal{T} = \{(\theta, \Psi(\theta) : \theta \in \mathbb{T}\}.$$

    b) $||\Psi - r_0||_{C^{n-1+\beta}} \leq KM^{-1}\kappa^{-1}\delta.$
    c) *The motion of $F$ restricted to $\mathcal{T}$ has rotation number $\omega$.*
    d) *If we denote by $g$ the map of the torus defined by*

$$F(\theta, \Psi(\theta)) = (g(\theta), \Psi(g(\theta))$$

    *and we denote by $h$ the map that conjugates $g$ to a rotation by $\omega$, i.e.*

$$g \circ h(\theta) = h(\theta + \omega)$$

    *normalized to $h(0) = 0$, we have*

$$(8.24) \qquad\qquad ||h - Id||_{C^{n-2+\beta}} \leq KM^{-1}\kappa^{-1}\delta.$$

*Moreover, if $n \geq 4$ we have for all the $F$ in a $C^{n+\beta}$ neighborhood of $F_0$:*

    e) *The mappings that associate $F$ to $\Psi$, $h$ respectively, are Lipschitz when we give $F$ the $C^{n+\beta}$ topology, $\Psi$ the $C^{n-2+\beta}$ topology and $h$ the $C^{n-3+\beta}$ topology.*
    f) *Define $\Gamma_F(\tilde{\Psi})$ the graph transform of $\tilde{\Psi}$ by:*

$$F(\mathrm{Graph}(\tilde{\Psi})) = \mathrm{Graph}(\Gamma_F(\tilde{\Psi})).$$

    *Assume that for some $C^{n-1+\beta}$ map we have*

$$||\tilde{\Psi} - \Gamma_F(\tilde{\Psi})||_{C^{n-1+\beta}} \leq \mu$$

    *and that $\tilde{\Psi}$ is in a $K\kappa^2$ neighborhood of a constant.*
    *Then, there is a $C^{n-1+\beta}$ function $\Psi^*$ whose graph is an invariant circle for $F$ such that*

$$||\tilde{\Psi} - \Psi^*||_{C^{n-1+\beta}} \leq K\kappa^{-1}\mu.$$

REMARK 8.20. For the experts in KAM theory, we call attention to the fact that Theorem 8.19 allows to conclude more regularity for the graph than for the conjugating function. Most of the versions of the KAM theorem study the conjugating function, hence, the regularity established for the curve is the same as that for the conjugating function. (Some theorems that establish that the graphs are more regular than the conjugacy are [**Sal04**], [**Pös82**].)

PROOF. We refer to [**Her83**] for the proof of the result. Here we only explain how the statement we have made follows from the statement in [**Her83**].

We note that the Theorem 5.4 of [**Her83**] is stated as a translated curve theorem for maps (not necessarily symplectic) of the form

$$(8.25) \qquad\qquad F(\theta, r) = (\theta + r + \alpha, r + \varphi(\theta, r)).$$

The proof of the translated curve result in [**Her83**] does not use any geometric feature (e.g. exact symplectic, intersection property) of the map considered.

Note that a twist mapping can be always put into the form (8.25) by a change of variables (possibly non-canonical) applying the implicit function theorem to define $r$ by the first component of (8.25). The map of the form (8.25) obtained through this procedure, has the same regularity of the original map. This change of variables does not change the regularity of the invariant circles or of the conjugacy of the motion on them to rotations.

Observe also that an exact symplectic mapping has the intersection property (the image of a nontrivial circle by the map has to intersect itself), and the intersection property of a map is preserved under any continuous change of variables. For any map with the intersection property, a translated curve has to be invariant.

The proof of Theorem 8.19 presented in [**Her83**] is extremely simple since it is not based on a rapidly convergent Nash-Moser method but rather on a Schauder-Tychonoff fixed point Theorem (when $n \geq 3$) or on a contraction mapping principle (when $n \geq 4$).

The fact that the proof uses a contraction theorem for an appropriate map is the reason why one has properties e) and f). Even if we will not use it here (and, hence not state the results) we note that using the results on composition in [**LO99**], it is possible to show that the operator considered in [**Her83**] is locally $C^\ell$ when the $F$ and the $\Psi$ are given in appropriate topologies (and $n$ is large enough) so that one gets that the invariant tori depend smoothly on $F$ in appropriate topologies. This justifies formal expansions in parameters.

□

PROOF OF THEOREM 8.12. Note that once that we have Theorem 8.19, Theorem 8.12 follows immediately if we consider the time-$2\pi$ map of the flow generated by the Hamiltonian (8.16). Note that if $\bar{k}(\mathcal{B}, \alpha, s; \varepsilon)$ is $C^{s+\beta}$, the time-$2\pi$ map is $C^{s-1+\beta}$. Moreover it is an exact symplectic twist map.

The time-$2\pi$ map corresponding to the integrable part of the Hamiltonian in (8.16) gives rise to a $C^{s+1+\beta}$ strongly integrable symplectic map. Given the form of the integrable part, the mapping will satisfy the twist condition.

By the standard dependence on parameters of differential equations, we obtain that the $C^{s-1+\beta}$ norms of the difference of the time-$2\pi$ maps can be bounded by $\varepsilon^{m+1}$.

Then, if we apply Theorem 8.19 to constant type numbers of Markov constant $\kappa = K\varepsilon^{(m+1)/2}$ with $n = s - 1$, and take into account Lemma 8.17 to control the spacing between such numbers, we obtain the statement of Theorem 8.12.                                                                    □

By Remark 8.11, applying Theorem 8.12 to the averaged Hamiltonian (8.16), when $r - 2m - 2 \geq 6$, and going back to the variables $(J, \varphi, s)$ using the change given by Theorem 8.9, we obtain the following result about the existence of invariant tori of Hamiltonian (8.4).

PROPOSITION 8.21. *Assume $r \geq 2m + 8$. Then, for $\varepsilon$ small enough, in any connected component of the non-resonant region $\mathcal{S}^L$ defined in (8.15), there exists a finite set of values $J_i$ such that:*

1) $\omega_i = J_i + \varepsilon\frac{\partial \bar{k}^{0,0}}{\partial J}(J_i, \varepsilon)$ *(see (8.16)) is a Diophantine number of constant type and Markov constant $K\varepsilon^{\frac{m+1}{2}}$.*
2) *There exists a torus $\mathcal{T}_i$, invariant by the flow of the Hamiltonian $k(J, \varphi, s; \varepsilon)$ given in (8.4), such that:*
    2.1) *The motion on the torus $\mathcal{T}_i$ is $C^1$-conjugated to a rigid translation of frequencies $(\omega_i, 1)$.*
    2.2) *The torus $\mathcal{T}_i$ can be written as a graph of the variable $J$ over the angle variables $(\varphi, s)$:*
    $$\mathcal{T}_i = \{(J, \varphi, s) \in \mathcal{S}^L, \ J = J_i + u_{\omega_i}(\varphi, s; \varepsilon)\},$$
    *where $u_{\omega_i}(\varphi, s; \varepsilon)$ is a $C^{r-2m-4-\eta}$ function, for any $\eta > 0$, and $\|u_{\omega_i}\|_{C^2} \leq$ cte.$\varepsilon$.*
3) *Denoting by*

(8.26)                  $$\mathbf{B}(\mathcal{A}, \rho) = \{\tilde{x} \in \tilde{\Lambda}_\varepsilon : \ \mathrm{dist}\,(\mathcal{A}, \tilde{x}) \leq \rho\},$$

   *for any $\mathcal{A} \subset \tilde{\Lambda}_\varepsilon$, one has that*

(8.27)                  $$\mathcal{S}^L \subset \bigcup_i \mathbf{B}(\mathcal{T}_i, K\varepsilon^{\frac{m+1}{2}}).$$

REMARK 8.22. Proposition 8.21 gives the primary KAM tori $\mathcal{T}_i$ in the variables $(J, \varphi, s)$, but we can obtain the tori in the original variables $(I, \varphi, s)$ using the change given by Proposition 8.2, which is $\varepsilon^2$-close to the identity. The tori thus obtained are invariant for the flow (3.4) and are of the form

$$\mathcal{T}_i = \{(I, \varphi, s) \in \mathcal{I} \times \mathbb{T}^2; I = I_i + U_{\omega_i}(\varphi, s; \varepsilon)\},$$

where the function $U_{\omega_i}$ verifies the same properties as $u_{\omega_i}$.

REMARK 8.23. The importance of Proposition 8.21 is that in the non-resonant region $\mathcal{S}^L$ we can find primary KAM tori with extremely small gaps between them.

It is important to note that, for a fixed value of $\varepsilon$, it is enough to consider a finite number of tori $\mathcal{T}_i$ to ensure that the regions $\mathbf{B}(\mathcal{T}_i, K\varepsilon^{\frac{m+1}{2}})$ cover all the non-resonant region $\mathcal{S}^L$. Of course, this number goes to infinity when $\varepsilon$ decreases to zero.

## 8.5. Analyzing the resonances

In this section we study the invariant sets close to resonant regions.

The goal is that we can cover the whole region with invariant objects (either primary tori, secondary tori or periodic orbits with (un)stable manifolds) at a distance less than $O(\varepsilon^{3/2})$.

The case of resonances of order 3 and higher will be studied in Section 8.5.1. It will not be different from the non-resonant region and will be enough to apply KAM theorem 8.12 to obtain primary tori with the required gaps.

The case of resonances or order 1 and 2 is significantly more involved and it will be done in Section 8.5.2, Section 8.5.3.

**8.5.1. Resonances of order 3 and higher.** In this section we study the reduced Hamiltonian $k(J, \varphi, s; \varepsilon)$ given in (8.4), in the regions close to the resonances of order 3 or bigger. To this end, we fix $j \geq 3$ and define:
$$(8.28) \qquad \mathcal{S}^{\mathcal{R}_j} = \bigcup_{-l_1/k_1 \in \mathcal{R}_j \backslash \mathcal{R}_1 \cup \cdots \cup \mathcal{R}_{j-1}} \{(J, \varphi, s) \in [-l_1/k_1 - L, -l_1/k_1 + L] \times \mathbb{T}^2\},$$

where $L$ is the constant provided in Theorem 8.9.

The connected components of this region are sets of the form:
$$\{(J, \varphi, s) \in [-l_1/k_1 - L, -l_1/k_1 + L] \times \mathbb{T}^2\}.$$

On them, after the averaging procedure given in Theorem 8.9, the Hamiltonian is a $\mathcal{C}^{r-2m-2}$ function of the form:
$$(8.29) \qquad \frac{\mathcal{B}^2}{2} + \varepsilon \bar{k}^{0,0}(\mathcal{B}; \varepsilon) + \varepsilon^j \left(U^{k_1, l_1}(k_1\alpha + l_1 s; \varepsilon) + \varepsilon^{m+1-j}\bar{k}^1(\mathcal{B}, \alpha, s; \varepsilon)\right).$$

We can apply Theorem 8.12 to Hamiltonian (8.29) since it is of the form (8.16) with $\varepsilon^j$ instead of $\varepsilon^{m+1}$. When we express the results in terms of the variables $(J, \varphi, s)$ using the change given by Theorem 8.9, we obtain:

PROPOSITION 8.24. *Assume $r \geq 2m + 8$. Then for any $3 \leq j \leq m$, in any connected component of the resonant region $\mathcal{S}^{\mathcal{R}_j}$, there exists a finite set of values $J_i$ such that:*

1) $\omega_i = J_i + \varepsilon \frac{\partial \bar{k}^{0,0}}{\partial J}(J_i, \varepsilon)$ *(see (8.29)) is a Diophantine number of constant type and Markov constant* $K\varepsilon^{j/2}$.

2) *There exists a torus* $\mathcal{T}_i$ *invariant by the flow of the Hamiltonian* $k(J, \varphi, s; \varepsilon)$ *given in (8.4), such that:*

    2.1) *The motion on the torus* $\mathcal{T}_i$ *can be* $\mathcal{C}^1$-*conjugated to a rigid translation of frequencies* $(\omega_i, 1)$.

    2.2) *The torus* $\mathcal{T}_i$ *can be written as a graph of the variable* $J$ *over the angle variables* $(\varphi, s)$:

$$\mathcal{T}_i = \{(J, \varphi, s) \in \mathcal{S}^{\mathcal{R}_j}, \ J = J_i + u_{\omega_i}(\varphi, s; \varepsilon)\}$$

*where* $u_{\omega_i}(\varphi, s; \varepsilon)$ *is a* $\mathcal{C}^{r-2m-4-\eta}$ *function, for any* $\eta > 0$, *and* $\|u_{\omega_i}\|_{\mathcal{C}^2} \leq \mathrm{cte.}\,\varepsilon$

3) *One has that*

(8.30)
$$\mathcal{S}^{\mathcal{R}_j} \subset \bigcup_i \mathbf{B}(\mathcal{T}_i, K\varepsilon^{j/2})$$

*where* $\mathbf{B}(\mathcal{A}, \delta)$ *is defined in (8.26).*

REMARK 8.25. Proposition 8.24 gives the primary KAM tori $\mathcal{T}_i$ in the variables $(J, \varphi, s)$, we can obtain the tori in the original variables $(I, \varphi, s)$ using the change given by Proposition 8.2 which is $\varepsilon^2$-close to the identity. The tori thus obtained are of the form

$$\mathcal{T}_i = \{(I, \varphi, s) \in \mathcal{I} \times \mathbb{T}^2; I = I_i + U_{\omega_i}(\varphi, s; \varepsilon)\},$$

where the function $U_{\omega_i}$ verifies the same properties as $u_{\omega_i}$.

**8.5.2. Preliminary analysis of resonances of order one or two.**
In this section we study in more detail the resonant regions $\mathcal{S}^{\mathcal{R}_j}$, defined in (8.28), for $j = 1, 2$, where $\mathcal{R}_1$ and $\mathcal{R}_2$ are defined in (5.4), (5.5).

This is the place where the standard mechanism of [**Arn64**] breaks down. Hence, this section is a place where the mechanism of diffusion presented in this paper differs from that of [**Arn64**].

As it is well known, there are easy examples where the KAM tori in regions $\mathcal{S}^{\mathcal{R}_j}$ have gaps of order $\varepsilon^{j/2}$. When $j = 1, 2$, these gaps are bigger than or comparable to $\varepsilon$ (which, as we will see later, is the size of the heteroclinic jumps). This is what has become known as the *large gap problem*.

What we will do in this and the next two sections is to show that, even if in the regions $\mathcal{S}^{\mathcal{R}_j}$ the primary KAM tori are rather scarce, we can find other geometric objects such as secondary KAM tori and stable and unstable manifolds of lower dimensional tori. These objects get rather close to the frontier of the gaps among the primary KAM tori.

From now on, we will work in one connected component of the domain $\mathcal{S}^{\mathcal{R}_j}$. More precisely, let us consider a resonance $-l_0/k_0 \in \mathcal{R}_j$, for $j = 1, 2$. In the component

$$\{(J, \varphi, s) \in [-l_0/k_0 - L, -l_0/k_0 + L] \times \mathbb{T}^2\},$$

the Hamiltonian $k(J, \varphi, s; \varepsilon)$ in (8.4) can be written in the averaged variables $(\mathcal{B}, \alpha, s)$ of Theorem 8.9 as:

$$
\bar{k}(\mathcal{B}, \alpha, s; \varepsilon) = \frac{\mathcal{B}^2}{2} + \varepsilon \bar{k}^{0,0}(\mathcal{B}; \varepsilon) + \varepsilon^j U^{k_0, l_0}(k_0 \alpha + l_0 s; \varepsilon)
$$

(8.31)

$$
+ \varepsilon^{m+1} \bar{k}^1(\mathcal{B}, \alpha, s; \varepsilon),
$$

defined, if we take $\varepsilon$ small enough, on

$$
\bar{D} := \{(\mathcal{B}, \alpha, s) \in \mathbb{R} \times \mathbb{T}^2, \quad |\mathcal{B} + l_0/k_0| \leq \bar{L}\},
$$

where $|L - \bar{L}| \leq \text{cte.}\,\varepsilon$, and where $\varepsilon \bar{k}^{0,0}$ is of class $\mathcal{C}^{r-2m}$, $U^{k_0, l_0}$ is a trigonometric polynomial in $(\varphi, s)$ and a polynomial in $\varepsilon$, and $\bar{k}^1$ is of class $\mathcal{C}^{r-2m-2}$ (see Remark 8.11).

Now, we can state explicitly the first part of the non-degeneracy conditions that constituted Hypothesis **H5** of Theorem 4.1:

**H5'** For any first or second order resonance, the function $U^{k_0, l_0}(\theta; 0)$ has a global maximum which is non-degenerate.

REMARK 8.26. By Theorem 8.9 applied to Hamiltonian (8.4) we have that, in the case of a first order resonance $-l_0/k_0 \in \mathcal{R}_1$

$$
U^{k_0, l_0}(\theta; 0) = \sum_{t=-N}^{N} \hat{k}^1_{tk_0, tl_0}(-l_0/k_0) e^{it\theta}
$$

where $\hat{k}^1_{k, l}$ are the Fourier coefficients of $k_1(J, \varphi, s)$, the second term in the Taylor expansion with respect to $\varepsilon$, of $k(J, \varphi, s; \varepsilon)$, the reduced Hamiltonian, which can be computed using formula (8.6). In the case of a second order resonance one can also obtain, using proposition 8.4, explicit formulas for $U^{k_0, l_0}(\theta; 0)$. Then, by Proposition 8.4, Hypothesis **H5'** can be checked by examining the term $h(p, q, I, \varphi, t; \varepsilon)$ in the original Hamiltonian (3.1).

The computations involved in these verifications are algebraic and quite explicit. In Chapter 13, we undertake the verification in a explicit example.

We note that the condition **H5'** is $C^j$, $j \geq 2$, open and dense in the space of polynomials $U^{k_0, l_0}(\theta, 0)$.

A closer look to the calculation of the coefficients (see 8.6) shows that the coefficient $\hat{h}_{tk_0, tl_0}$ of the perturbation (4.2) enters linearly in the expression for $U^{k_0, l_0}(\theta, 0)$. Hence, we conclude that, in the space of original Hamiltonians restricted to the manifold $\tilde{\Lambda}_\varepsilon$, the hypothesis **H5'** holds in a $C^j$ $j \geq 2$ open and dense of Hamiltonians.

To study Hamiltonian (8.31) in the set $\bar{D}$, we consider the change of variables depending on time given by

(8.32)            $b = k_0(\mathcal{B} + l_0/k_0), \quad \theta = k_0 \alpha + l_0 s, \quad s = s.$

The change (8.32) is not a true symplectic change of variables but it is conformally symplectic, hence the new system of differential equations verified by $(b, \theta, s)$ is also Hamiltonian, of Hamiltonian:

$$(8.33) \qquad \bar{K}(b, \theta, s; \varepsilon) = \bar{K}^0(b; \varepsilon) + \varepsilon^j \bar{V}(\theta; \varepsilon) + \varepsilon^{m+1} \bar{K}^1(b, \theta, s; \varepsilon),$$

with

$$(8.34) \qquad \begin{aligned} \bar{K}^0(b, \varepsilon) &= \frac{b^2}{2} + \varepsilon k_0^2 \bar{k}^{0,0}(-l_0/k_0 + b/k_0; \varepsilon), \\ \bar{V}(\theta; \varepsilon) &= k_0^2 U^{k_0, l_0}(\theta; \varepsilon), \\ \bar{K}^1(b, \theta, s; \varepsilon) &= k_0^2 \bar{k}^1 \left( -l_0/k_0 + b/k_0, \frac{\theta - l_0 s}{k_0}, s; \varepsilon \right). \end{aligned}$$

Note that $\bar{K}^0$ is of class $C^{r-2m}$, $\bar{V}$ is analytic, and $\bar{K}^1$ is of class $C^{r-2m-2}$.

Up to order $\varepsilon^m$, the Hamiltonian (8.33) is given by $\bar{K}_0(b; \varepsilon) + \varepsilon^j \bar{V}(\theta; \varepsilon)$ which is a one degree of freedom Hamiltonian. Moreover, it is close to a pendulum-like Hamiltonian

$$\frac{b^2}{2} + \varepsilon^j \bar{V}(\theta; 0).$$

By hypothesis **H5'**, this Hamiltonian has a hyperbolic saddle at $(0, \theta_1)$, where $\theta_1$ is the maximum of $U^{k_0, l_0}(\theta; 0)$. The eigenvalues of the linearization are, of course, $\pm \varepsilon^{j/2} \sqrt{-V''(\theta_1)} + o(\varepsilon^{j/2})$.

Using the implicit function theorem, one can see easily that the Hamiltonian $\bar{K}_0(b; \varepsilon) + \varepsilon^j \bar{V}(\theta; \varepsilon)$ given in (8.33) also has a saddle at $(b(\varepsilon), \theta_1(\varepsilon)) = (0, \theta_1) + O(\varepsilon)$. This saddle corresponds to a hyperbolic periodic orbit if we also include the $s$ variable as the time coordinate. The function $b(\varepsilon)$ is of class $C^{r-2m-1}$, and $\theta_1(\varepsilon)$ is analytic.

To analyze the behavior of this pendulum-like system, we will find it convenient to make the translation

$$(8.35) \qquad y = b - b(\varepsilon), \qquad x = \theta - \theta_1(\varepsilon), \qquad s = s,$$

obtaining the $C^{r-2m-2}$ Hamiltonian

$$(8.36) \qquad K(y, x, s; \varepsilon) = h^0(y; \varepsilon) + \varepsilon^j U(x; \varepsilon) + \varepsilon^{m+1} S(y, x, s; \varepsilon)$$

where

$$(8.37) \qquad \begin{aligned} h^0(y; \varepsilon) &= \bar{K}^0(y + b(\varepsilon); \varepsilon) - \bar{K}^0(b(\varepsilon); \varepsilon) \\ U(x; \varepsilon) &= \bar{V}(x + \theta_1(\varepsilon); \varepsilon) - \bar{V}(\theta_1(\varepsilon); \varepsilon) \\ S(y, x; \varepsilon) &= \bar{K}^1(y + b(\varepsilon), x + \theta_1(\varepsilon), s; \varepsilon) \end{aligned}$$

where we have subtracted a constant term, the energy of the saddle, to normalize:

$$(8.38) \quad h^0(0; \varepsilon) = \frac{\partial h^0}{\partial y}(0; \varepsilon) = 0, \ U(0; \varepsilon) = \frac{\partial U}{\partial x}(0; \varepsilon) = 0, \ \frac{\partial^2 U}{\partial x^2}(0; \varepsilon) < 0,$$

and $x = 0$ is a global maximum of $U$.

Let us note for further reference that the expression of $h^0(y;\varepsilon)$ given in (8.37) shows that $h^0(y;\varepsilon)$ is a $C^{r-2m-1}$ function verifying:

$$(8.39) \qquad h^0(y;\varepsilon) = \frac{y^2}{2}\hat{h}(y;\varepsilon) = \frac{y^2}{2}(1+\varepsilon\tilde{h}(y;\varepsilon))$$

where $\varepsilon\tilde{h}$ is a $C^{r-2m-2}$ function.

Notice that the Hamiltonian $K(y,x,s;\varepsilon)$ is a $2\pi k_0$-periodic in $s$ function defined in the domain $D_{k_0}$ given by:

$$(8.40) \qquad D_{k_0} = \{(y,x,s) \in \mathbb{R} \times (\mathbb{R}/(2\pi k_0\mathbb{Z}))^2, \quad |y| \le k_0\bar{L}\}.$$

From now on, we work in the variables $(y,x,s)$, and, when necessary, we will recover the original Hamiltonian by performing the changes of variables (8.32), (8.35).

The first important point is that, up to order $\varepsilon^m$, the Hamiltonian (8.36) is given by the one degree of freedom $C^{r-2m-1}$ Hamiltonian:

$$(8.41) \qquad K_0(y,x;\varepsilon) = h^0(y;\varepsilon) + \varepsilon^j U(x;\varepsilon),$$

and the energy level $K_0(y,x;\varepsilon) = 0$ consists on the saddle $(0,0)$ and its separatrices.

REMARK 8.27. Note that the $2\pi k_0$-periodic Hamiltonians (8.33) and (8.36) are, up to order $\varepsilon^m$, $2\pi$-periodic. This is the well known effect that is colloquially described as saying that the resonance has "$k_0$ eyes" (see Figure 8.1).

**8.5.3. Primary and secondary tori near the first and second order resonances.** As the Hamiltonian $K_0(y,x;\varepsilon)$ given in (8.41) is $2\pi$-periodic, the region $D_{k_0}$ given in (8.40) can be seen as $k_0$ copies of the region

$$(8.42) \qquad D = \{(y,x,s) \in \mathbb{R} \times \mathbb{T}^2, \quad |y| \le k_0\bar{L}\}.$$

where this Hamiltonian is well defined.

The region $D$—and then $D_{k_0}$—is filled by the energy surfaces of the Hamiltonian $K_0$, given in (8.41):

$$\mathcal{T}_E^0 = \{(y,x,s) \in [-k_0\bar{L}, k_0\bar{L}] \times \mathbb{T}^2 : K_0(y,x;\varepsilon) = E\}.$$

$\mathcal{T}_E^0$ will, of course, be invariant by the Hamiltonian flow of $K_0$.

The energy surfaces that correspond to values of $E > 0$, are two primary tori in the sense of Definition 2.2 in Section 2.1. These primary tori can be written as a graph of the variable $y$ over the angular variables $(x,s)$.

The energy surface corresponding to $E = 0$ consists of the saddle $(0,0)$ and the homoclinic orbits to it. We will refer to $\mathcal{T}_0^0$ as the separatrix loop.

When $E < 0$ the invariant surfaces (which are contained inside the region bounded by the separatrix loop $\mathcal{T}_0^0$) are tori of different topology than the primary tori, since they are contractible to a point. They are secondary tori in the sense of Definition 2.2 and cover all the region inside the separatrix loop. When $E < 0$ is close to zero, the curve $\mathcal{T}_E^0$ will be very close to the separatrix loop. These circles that get very close to the separatrix loop will allow us to transverse the resonance.

The main point of this section will be to show that many of the invariant tori of $K_0(y, x; \varepsilon)$ (both primary and secondary) survive when we add the perturbation term $\varepsilon^{m+1} S(y, x, s; \varepsilon)$ in (8.36). We will pay special attention to getting some of these tori close—$O(\varepsilon^{3/2})$ will be enough for our purposes— to the separatrix loop of $K_0$. (See Figure 8.1.)

To establish this, we will put the unperturbed Hamiltonian $K_0(y, x; \varepsilon)$ in action-angle variables and apply KAM theorem 8.19 to the time-$2\pi k_0$ map of the Hamiltonian (8.36).

As it is well known, systems with hyperbolic critical points and separatrices do not have global action-angle variables, because the angle variable is not defined in the separatrix, where the frequency of the motion becomes zero.

We will need to use different action-angle variables inside and outside the separatrix. The main difficulty arises because the changes of variables reducing to action-angle become singular as we approach the separatrix. Nevertheless, we will control how singular these changes become in terms of the distance to the separatrix loop. Hence, using that the perturbation $S$ is small—it contains a factor $\varepsilon^{m+1}$ and $m$ is large—we will see that it remains small when written in the action angle-variables provided that we do not consider it too close to the separatrix. Then, we can apply the KAM theorem 8.12 and can obtain tori close to a high power of $\varepsilon$ to the separatrix loop. For subsequent purposes getting tori which are at a distance $\varepsilon^{3/2}$ is enough.

REMARK 8.28. Similar arguments for analytic systems were used in [Neĭ84]. In the analytic case considered there, the tori are exponentially close to the separatrix. Since our systems are only finitely differentiable we need to develop arguments different from those in [Neĭ84]. Since we can only perform a finite number of averaging transformations, we only obtain that there are invariant tori at a distance to the separatrix not larger than a power of the perturbation.

REMARK 8.29. We will find it convenient to consider three different regions in the domain $D_{k_0}$ in which the behavior of the tori is very different.

To studying the tori in different regions in different ways is justified because of the fact that the tori have different quantitative properties. Therefore, the leading terms in asymptotic expansions take different forms.

Tori far away from the resonance can be considered extremely flat, whereas tori very close to the resonance seem to be bunched near the critical point. This can be understood by looking at the level sets of the Hamiltonian $H(y, x) = \frac{y^2}{2} + \cos x - 1$.

When $E \gg 1$, we have that $\mathcal{T}_E$, $\mathcal{T}_{E+\delta}$ are graphs of functions which are at distance $O(\delta)$ at all the points. Nevertheless we see that $\mathcal{T}_\delta$ and $\mathcal{T}_0$ are two tori such that for $x = \pi/2$ the distance is $O(\delta)$, whereas at $x = 0$ the distance is $O(\sqrt{\delta})$.

For small $E$, the same situation persists. The tori $\mathcal{T}_E$ and $\mathcal{T}_{E+\delta}$ are at a distance $O(\delta)$ in the middle and at a distance $\frac{\delta}{\sqrt{E}}$ near the singularity $x = 0$.

It is clear that if we are considering the tori at distances $\delta = \varepsilon^{3/2}$ and are interested in effects of size $\varepsilon$, we have to distinguish the size of their energies. This will affect which terms in an expansion are dominant.

The choices we have made below are not the only possible ones, but the above considerations show that there are quantitatively different features in different regions.

$D_{k_0}$ will be divided in three regions. $D_f$ is the region far from the separatrix, $D_o$ close to the separatrices but out of the region bounded by the separatrix loop and $D_{in}$ close to the separatrices but inside the separatrix loop.

The precise definitions we have found useful are:

$$(8.43)\, D_f \;=\; \{(y,x,s) \in D_{k_0} : K_0(y,x;\varepsilon) = E,\ c_1\varepsilon^j \le E \le c_2\bar{L}\}$$

$$(8.44)\, D_o \;=\; \{(y,x,s) \in D_{k_0} : K_0(y,x;\varepsilon) = F,\ c_3\varepsilon^\alpha \le F \le c_1\varepsilon^j\}$$

$$(8.45)\, D_{in} \;=\; \{(y,x,s) \in D_{k_0} : K_0(y,x;\varepsilon) = G,\ -c_4\varepsilon^j \le G \le -c_3\varepsilon^\alpha\}$$

where $\alpha$, for the time being, is arbitrary provided that $\alpha > j = 1, 2$, which is the order of the resonance, and $c_i$ are suitable constants independent of $\varepsilon$. (In particular, $c_4 < c$, where $-c$ is the minimum of $U$, is taken small enough in order that $D_{in}$ does not contain any other critical point of $K_0$.)

As we will see, $\alpha$ controls how close to the separatrices we claim to find tori. Roughly speaking, we can take $\alpha$ any number, provided that we are willing to perform enough steps of averaging (this amounts to taking $m$ big enough in (8.36), which can be done just by assuming that the original system is differentiable enough).

Even if we state Theorem 8.30 for arbitrarily $\alpha > j$ and $m$ large enough with respect to $\alpha$, we note that for the subsequent applications in this paper, it will suffice to take any value $\alpha > 1+j/2$, $j = 1, 2$. Later, in Corollary 8.31, for the sake of definiteness, we will take $\alpha = 3/2 + j/2$, and $m = 26$.

The following Theorem 8.30 establishes the existence of primary tori in $D_f \cup D_o$ and secondary tori in $D_{in}$.

The three regions (8.43), (8.44) and (8.45) cover the resonant region, except for a thin neighborhood of the separatrix given by

$$(8.46) \qquad\qquad \{(y,x,s) \in D_{k_0} : |K_0(y,x;\varepsilon)| \le c_3\varepsilon^\alpha\},$$

and a small neighborhood of the center of the resonant region, given by $K_0(y,x;\varepsilon) \le -c_4\varepsilon^j$. The region (8.46) not covered in the present analysis contains what in physical language is called the "chaotic sea". In Section 8.5.5 we will identify features, other than invariant tori, in (8.46) namely, periodic orbits and their (un)stable manifolds.

Recall that $K_0(y, x; \varepsilon)$ as given in (8.41) describes a pendulum. Since have assumed that we do not have any other critical points in the region we consider, all the orbits we will study are periodic orbits.

Since we will be applying KAM arguments, the values of the frequency $\omega(E)$ and its non-degeneracy properties will play an important role.

An orbit of $K_0$ of energy $E$ has frequency

$$(8.47) \qquad \omega(E) = \frac{2\pi}{T(E)}, \ T(E) = \int_{K_0^{-1}(E)} \frac{dx}{y}.$$

THEOREM 8.30. *Consider the $\mathcal{C}^{r-2m-2}$ reduced Hamiltonian $K(y, x, s; \varepsilon)$ as in (8.36), inside the region $D_{k_0}$ given in (8.40). Consider $\alpha > j$, for $j = 1, 2$, $m \geq \max(11j - 1, 14(\alpha - j) - 1 - j/2)$, and assume that $r \geq 2m + 8$. Then, for $|\varepsilon|$ small enough, one has:*

1) ***Primary tori far from resonance.*** *There exists a set of values $E_1 < \cdots < E_{l_E}$ verifying $c_1 \varepsilon^j \leq E_i \leq c_2 \bar{L}$, such that:*

   1.1) *The frequencies $\omega(E_i)$ are Diophantine numbers of constant type and Markov constant $K \varepsilon^{\frac{m+1-6j}{2}}$ (see Definition 8.16).*

   1.2) *For any value $E_i$, there exist two primary invariant tori $\mathcal{T}_{E_i}^{\pm}$ of Hamiltonian (8.36) contained in $D_f$.*

   1.3) *The motion on the tori $\mathcal{T}_{E_i}^{\pm}$ is $\mathcal{C}^1$-conjugated to a rigid translation of frequencies $(\omega(E_i), 1)$.*

   1.4) *These tori can be written as:*

   $$\mathcal{T}_{E_i}^+ = \{(y, x, s) \in D_f, \ K_{E_i}(y, x, s; \varepsilon) = E_i, \ y > 0\},$$
   $$\mathcal{T}_{E_i}^- = \{(y, x, s) \in D_f, \ K_{E_i}(y, x, s; \varepsilon) = E_i, \ y < 0\},$$

   *where $K_{E_i}(y, x, s; \varepsilon)$ is a $\mathcal{C}^{r-2m-5-\eta}$ function, for any $\eta > 0$, given by*

   $$(8.48) \qquad K_{E_i}(y, x, s; \varepsilon) = K_0(y, x; \varepsilon) + O_{\mathcal{C}^2}\left(\varepsilon^{\frac{m+1-11j}{2}}\right).$$

   1.5) $D_f \subset \bigcup_i B\left(\mathcal{T}_{E_i}^{\pm}, \varepsilon^{\frac{m+1-7j}{2}}\right)$, *where*

   $$(8.49) \qquad B\left(\mathcal{T}_E^{\pm}, \delta\right) = \{(y, x, s) \in D_{k_0}, \ |K_0(y, x; \varepsilon) - E| \leq \delta\}.$$

2) ***Primary tori close to resonance.*** *There exists a set of values $F_1 < \cdots < F_{l_F}$ verifying $c_3 \varepsilon^{\alpha} \leq F_i \leq c_1 \varepsilon^j$, such that:*

   2.1) *The frequencies $\omega(F_i)$ are Diophantine numbers of constant type and Markov constant $K \varepsilon^{\frac{m+1-6(\alpha-j)-j/2}{2}}$.*

   2.2) *For any value $F_i$, there exist two primary invariant tori $\mathcal{T}_{F_i}^{\pm}$ of Hamiltonian (8.36) contained in $D_o$.*

   2.3) *The motion on the tori $\mathcal{T}_{F_i}^{\pm}$ is $\mathcal{C}^1$-conjugated to a rigid translation of frequencies $(\omega(F_i), 1)$.*

2.4) *These tori can be written as:*

$$\mathcal{T}_{F_i}^+ = \{(y, x, s) \in D_o, \ K_{F_i}(y, x, s; \varepsilon) = F_i, \ y > 0\}$$
$$\mathcal{T}_{F_i}^- = \{(y, x, s) \in D_o, \ K_{F_i}(y, x, s; \varepsilon) = F_i, \ y < 0\}$$

*where $K_{F_i}(y, x, s; \varepsilon)$ is a $\mathcal{C}^{r-2m-5-\eta}$ function, for any $\eta > 0$, given by:*

(8.50) $$K_{F_i}(y, x, s; \varepsilon) = K_0(y, x; \varepsilon) + O_{\mathcal{C}^2}\left(\varepsilon^{\frac{m+1+j/2-14(\alpha-j)}{2}}\right).$$

2.5) $D_o \subset \bigcup_i B\left(\mathcal{T}_{F_i}^\pm, \varepsilon^{\frac{m+1+j/2-10(\alpha-j)}{2}}\right).$

3) ***Secondary tori close to resonance.*** *There exists a set of values $G_1 < \cdots < G_{l_G}$ verifying $-c_4\varepsilon^j \leq G_i \leq -c_3\varepsilon^\alpha$, such that:*

3.1) *The frequencies $\omega(G_i)$ are Diophantine numbers of constant type and Markov constant $K\varepsilon^{\frac{m+1-6(\alpha-j)-j/2}{2}}$.*

3.2) *For any value $G_i$, there exists a secondary invariant torus $\mathcal{T}_{G_i}$ of Hamiltonian (8.36) contained in $D_{in}$, contractible to the set*

$$\{(0, a, s), a \in \mathbb{R}, \ s \in \mathbb{R}/(2\pi k_0 \mathbb{Z})\} \subset D_{in}.$$

3.3) *The motion on the torus $\mathcal{T}_{G_i}$ is $\mathcal{C}^1$-conjugated to a rigid translation of frequencies $(\omega(G_i), 1)$.*

3.4) *This torus can be written as:*

$$\mathcal{T}_{G_i} = \{(y, x, s) \in D_{in}, \ K_{G_i}(y, x, s; \varepsilon) = G_i\}$$

*where $K_{G_i}(y, x, s; \varepsilon)$ is a $\mathcal{C}^{r-2m-5-\eta}$ function, for any $\eta > 0$, given by:*

(8.51) $$K_{G_i}(y, x, s; \varepsilon) = K_0(y, x; \varepsilon) + O_{\mathcal{C}^2}\left(\varepsilon^{\frac{m+1+j/2-14(\alpha-j)}{2}}\right).$$

3.5) $D_{in} \subset \bigcup_i B\left(\mathcal{T}_{G_i}, \varepsilon^{\frac{m+1+j/2-10(\alpha-j)}{2}}\right).$

Theorem 8.30 gives the existence of invariant tori in $D_{k_0}$. In the following Corollary 8.31, we will make more explicit assertions about the proximity of these tori. We will also make precise assertions of their properties when expressed as graphs of functions from the angle variables to the action variable. The proof of Corollary 8.31 from Theorem 8.30 is just an easy application of the implicit function theorem. Nevertheless, we state it explicitly because the properties of the tori as graphs will be useful later in Chapter 9 when we study how the tori behave under the scattering map.

COROLLARY 8.31. *Let $K(y, x, s; \varepsilon)$ be the reduced Hamiltonian function as in (8.36), inside the region $D_{k_0}$ given in (8.40). Consider $\alpha = j/2 + 3/2$, for $j = 1, 2$, and $m \geq 26$. Then, if $r \geq 2m + 8$, the tori defined in Theorem 8.30 verify:*

(1) *For any value $E_i$, the primary tori $\mathcal{T}_{E_i}^\pm$ can be written as graphs of the action $y$ over the angles $(x, s)$:*

$$\mathcal{T}_{E_i}^\pm = \{(y, x, s) \in D_f, \ y = f_{E_i}^\pm(x, s; \varepsilon)\}.$$

(2) *For any value $F_i$, the primary tori $\mathcal{T}_{F_i}^{\pm}$ can be written as graphs of the action $y$ over the angles $(x, s)$:*

$$\mathcal{T}_{F_i}^{\pm} = \{(y, x, s) \in D_o, \ y = f_{F_i}^{\pm}(x, s; \varepsilon)\}.$$

(3) *There exists $\rho_0 > 0$ such that for any $0 < \rho_0 \leq \rho < \pi$, and for any value $G_i$, each of the components of*

$$\mathcal{T}_{G_i} \cap \{(y, x, s) : \ x \in I_\rho\}, \quad I_\rho = \cup_{l=0}^{k_0-1}[2\pi l + \rho, 2\pi(l + 1) - \rho]\},$$

*that we will denote by $\mathcal{T}_{G_i}^{\pm, \rho}$, can be written as a graph of the action $y$ over the angles $(x, s)$:*

$$\mathcal{T}_{G_i}^{\pm, \rho} = \{(y, x, s) \in D_{in}, x \in I_\rho, \ y = f_{G_i}^{\pm}(x, s; \varepsilon)\}$$

(4) *There exists a constant $K$, independent of $\varepsilon$, such that*

$$
\begin{aligned}
|E_i - E_{i+1}| &\leq K\varepsilon^{(m+1-7j)/2} \leq K\varepsilon^{\frac{3}{2}+\frac{j}{2}} \\
|F_i - F_{i+1}| &\leq K\varepsilon^{(m+1+j/2-10(\alpha-j))/2} \leq K\varepsilon^{\frac{3}{2}+\frac{j}{2}} \\
|G_i - G_{i+1}| &\leq K\varepsilon^{(m+1+j/2-10(\alpha-j))/2} \leq K\varepsilon^{\frac{3}{2}+\frac{j}{2}}
\end{aligned}
$$

(5)

$$
\begin{aligned}
|E_1 - F_{l_F}| &\leq K\varepsilon^{(m+1-7j)/2} \leq K\varepsilon^{\frac{3}{2}+\frac{j}{2}} \\
|F_1 - G_{l_G}| &\leq K\varepsilon^{\alpha} + \varepsilon^{(m+1+j/2-10(\alpha-j))/2} \leq K\varepsilon^{\frac{3}{2}+\frac{j}{2}}
\end{aligned}
$$

(6) *All these functions $f_v = f_v^{\pm}$ are at least of class $\mathcal{C}^2$, and, denoting by $D$ the derivatives with respect to $x$ and $s$, for $v = E_i$, $i = 1, \ldots, l_E$, $v = F_i$, $i = 1, \ldots, l_F$, and $v = G_i$, $i = 1, \ldots, l_G$, they verify:*

   (a) *There exists a function $\mathcal{Y}(x, E)$, given explicitly in (8.56), such that:*

$$
\begin{aligned}
|f_{E_i} - \mathcal{Y}(x, E_i)|_{\mathcal{C}^1} &\leq K\varepsilon^{(m+1-12j)/2} \leq K\varepsilon^{3/2} \\
\text{(8.52)} \quad |f_{F_i} - \mathcal{Y}(x, F_i)|_{\mathcal{C}^1} &\leq K\varepsilon^{(m+1-j/2-14(\alpha-j))/2} \leq K\varepsilon^{3/2} \\
|f_{G_i} - \mathcal{Y}(x, G_i)|_{\mathcal{C}^1} &\leq K\varepsilon^{(m+1-j/2-14(\alpha-j))/2} \leq K\varepsilon^{3/2}
\end{aligned}
$$

   (b) $|Df_v| \leq K\varepsilon^{j/2}$, $|D^2 f_v| \leq K\varepsilon^{j/2}$.

   (c) *For any two consecutive values $v$ and $\bar{v}$ we have:*

$$\text{(8.53)} \qquad |f_v - f_{\bar{v}}|_{\mathcal{C}^1} \leq \frac{|v - \bar{v}|}{\varepsilon^{j/2}} \leq K\varepsilon^{3/2}.$$

REMARK 8.32. If we go back to the original variables $(I, \varphi, s)$ through the changes given by Proposition 8.2, Theorem 8.9 and the changes (8.32), and (8.35), we obtain that the tori inside the region $\mathcal{S}^{\mathcal{R}^j}$, $J = 1, 2$ are given by,

$$I = -k_0/l_0 + U_v^{\pm}(\varphi, s; \varepsilon)$$

where the functions $U_v^{\pm}(\varphi, s; \varepsilon)$ verify the same properties than the functions $f_v^{\pm}$. This is, of course, a reflection of the fact that the properties of proximity between the tori are geometric properties that are invariant under changes of variables.

**8.5.4. Proof of Theorem 8.30 and Corollary 8.31.** As indicated, the main technique will be to estimate the singularities of the change to action angle variables in the curves that are close to the separatrix. As it is well known, these action angle variables are expressed as integrals of a form over the energy surfaces. The main source of problems is that the surfaces become singular as we approach the separatrix.

REMARK 8.33. An exercise that the authors found useful is to go over the calculations of the action angle variables of the physical pendulum. In such a case, the integrals appearing in the definition of the action variables are explicitly expressed in terms of elliptic functions, so that the order of magnitude of the singularities can be checked.

In the following Lemma 8.34, we want to fix $\varepsilon$ sufficiently small, and $0 < \delta \leq 1$, and consider the foliation given by the level sets

$$(8.54) \qquad\qquad h(y; \varepsilon) + \delta U(x; \varepsilon) = E,$$

and obtain that—excluding a small interval—we can consider them as graphs.

Furthermore, we want to obtain bounds on the singularities which are uniform in $\varepsilon$, $\delta$.

The idea of the proof is very simple. We just need to realize that $h(y; \varepsilon) + \delta U(x; \varepsilon) \simeq \frac{y^2}{2} + \delta U(x; \varepsilon)$, so that the main term of the solution of (8.54) is

$$(8.55) \qquad\qquad y = \pm\sqrt{2(E - \delta U(x; \varepsilon))}.$$

Indeed, if we express the $y$ in (8.54) as a function of (8.55), equation (8.54) becomes an equation which can be dealt by the implicit function theorem. Note that the energy levels of the pendulum, when expressed as a graph, have square root singularities at $E = 0$.

What we want to show is that, if we add extra terms to the Hamiltonian, the level sets can be still expressed as graphs and that the singularities are still the square root singularities. This will be important for us later since the singularities of the graph control the singularities of the action-angle variables.

As it is standard in dynamical systems, when dealing with perturbations of systems that involve singularities, it is more convenient to formulate the problem in a system of coordinates which incorporates the singularities. Hence we will express the results as a function of (8.55) so that the resulting functions are smooth.

LEMMA 8.34. *Let $U(x; \varepsilon)$ and $h(y; \varepsilon)$ be two $C^n$ functions, $n = r - 2m - 1$, defined in $(y, x) \in \mathbb{R} \times \mathbb{T}$. Assume:*

(1) *$U(x; \varepsilon)$ has a non-degenerate global maximum at the origin.*

(2) $U(x;\varepsilon)$ *verifies* $-c \leq U(x;\varepsilon) \leq 0$, *and* $U(x;\varepsilon) = -ax^2 + O(x^3)$, *as* $x \to 0$.

(3) $h(y;\varepsilon)$ *is of the form* (8.39).

*Let* $\delta \in [-\delta_0, \delta_0]$ *and consider the equation* (8.54). *Then*

(1) *For* $|\varepsilon|$ *small enough, and for* $-c\delta \leq E \leq M$, *where* $M$ *is independent of* $\varepsilon$ *and* $\delta$, *equation* (8.54) *defines two functions* $y = \mathcal{Y}_\pm(x, E)$ *on* $\mathcal{I} = \mathcal{I}(\varepsilon, \delta) := \{(x, E), x \in \mathbb{T}, E - \delta U(x;\varepsilon) \geq 0\}$ *such that:*

(8.56) $$\mathcal{Y}_\pm(x, E) = \pm(1 + \varepsilon b)\ell(x, E) + \varepsilon\tilde{\mathcal{Y}}_\pm(\ell(x, E)),$$

*where* $\ell(x, E) = \sqrt{2(E - \delta U(x;\varepsilon))}$, *and:*
  (a) $b$ *is independent of* $\delta$, *and* $\tilde{\mathcal{Y}}_\pm(0) = \tilde{\mathcal{Y}}'_\pm(0) = 0$.
  (b) $\varepsilon\tilde{\mathcal{Y}}_\pm$ *is a* $\mathcal{C}^n$ *function and*

$$\|\varepsilon\tilde{\mathcal{Y}}_\pm \circ \ell\|_{\mathcal{C}^s(\mathcal{I}_{E_0})} \leq \text{cte}.\,\varepsilon, \quad s = 0, 1,$$

(8.57)

$$\|\varepsilon\tilde{\mathcal{Y}}_\pm \circ \ell\|_{\mathcal{C}^s(\mathcal{I}_{E_0})} \leq \frac{\text{cte}.\,\varepsilon}{E_0^{s-1/2}}, \quad 2 \leq s \leq n.$$

*where* $\mathcal{I}_{E_0} := \{(x, E), x \in \mathbb{T}, E \geq E_0 > 0\}$

(2) *There exists* $\rho$ *independent of* $\varepsilon$ *and* $\delta$, *such that for* $-\rho \leq x \leq \rho$, *equation* (8.54) *defines two functions* $x = \mathcal{X}_\pm(y, E)$ *on* $\mathcal{J} = \mathcal{J}(\varepsilon, \delta) := \{(y, E), U(\rho)\delta < E < 0, y \in \mathbb{R}, h(y;\varepsilon) - E < -\delta U(\rho;\varepsilon)\}$ *such that:*

(8.58) $$x = \mathcal{X}_\pm(y; E) = m(y, E) + \tilde{\mathcal{X}}_\pm(m(y, E)),$$

*where* $m(y, E) = \sqrt{\frac{1}{a\delta}(h(y;\varepsilon) - E)}$, *and:*
  (a) $\tilde{\mathcal{X}}_\pm(0) = \tilde{\mathcal{X}}'_\pm(0) = 0$.
  (b) $\tilde{\mathcal{X}}_\pm(t)$ *is a* $\mathcal{C}^n$ *function and*

(8.59) $$\|\tilde{\mathcal{X}}_\pm \circ m\|_{\mathcal{C}^s(\mathcal{J}_{E_0})} \leq \frac{\text{cte}.}{|E_0|^{s-3/2}}, \quad 2 \leq s \leq n,$$

*where* $\mathcal{J}_{E_0} := \{(y, E), U(\rho)\delta < E \leq E_0 < 0, y \in \mathbb{R}, h(y;\varepsilon) - E < -\delta U(\rho;\varepsilon)\}$, *and its* $\mathcal{C}^0$ *and* $\mathcal{C}^1$ *norms are bounded independently of* $\delta$ *and* $\varepsilon$.

PROOF. By (8.39) equation (8.54) is equivalent to

$$\frac{y^2}{2}(1 + \varepsilon\tilde{h}(y;\varepsilon)) = \frac{t^2}{2},$$

where $\pm t = \sqrt{2(E - \delta U(x;\varepsilon))} := \ell(x; E)$. Writing $y = tz$, the equation above becomes

$$\frac{z^2}{2} + \varepsilon z^2 \tilde{h}(tz;\varepsilon) = 1.$$

$z = 1$ is a solution when $\varepsilon = 0$. The implicit function theorem gives us the existence of a solution of the form

$$z = 1 + \varepsilon g(t;\varepsilon) = 1 + \varepsilon b(\varepsilon) + \varepsilon\tilde{g}(t;\varepsilon),$$

where $b(\varepsilon) = g(0; \varepsilon)$, so that $\varepsilon \tilde{g}$ is $\mathcal{C}^n$ function with a bounded $\mathcal{C}^n$ norm verifying $\tilde{g}(0, \varepsilon) = 0$. Therefore, the solution of equation (8.54) is given by

$$y = \pm \ell(x, E)(1 + \varepsilon b) + \varepsilon \ell(x, E) \tilde{g}(\ell(x, E); \varepsilon).$$

Writing $\tilde{y}(t) = t\tilde{g}(t; \varepsilon)$ we obtain (8.56). Bounds (8.57) follow from Faa-di-Bruno formulae.

The proof of part (2) of the Lemma is analogous.

$\square$

The next Proposition 8.35 studies the action-angle variables $(A, \psi)$ associated to the Hamiltonian $K_0(y, x; \varepsilon)$ in the domain $D_f$ defined in (8.43).

In order to simplify the notation, we note that $K_0$ is independent of $s$ and that it is $2\pi$-periodic in $x$ rather than $2k_0\pi$.

Therefore, we will consider the domain

$$D_f^* = \{(y, x) \in \mathbb{R} \times \mathbb{T}, \ \exists\, l \in \{0, 1, \cdots, k_0\}, \ (y, x + 2\pi l, s) \in D_f\}$$

in the variables $y, x$, which is $2\pi$-periodic.

Roughly speaking, what we do is to restrict the action-angle variables in one of the "eyes of the resonance". The results obtained in one eye extend to the other eyes by the $2\pi$-periodicity in $x$ of $K_0$, and clearly are uniform for all $s$ since $s$ does not enter in the Hamiltonian.

PROPOSITION 8.35. *Consider a Hamiltonian $K_0(y, x; \varepsilon)$ as in (8.41) of class $\mathcal{C}^{r-2m-1}$, with $h^0$ verifying (8.39), in the region $D_f^*$. Then, there exists a $\mathcal{C}^{r-2m-2}$ change of variables*

(8.60)                              $\chi_f : \tilde{D}_f \ \rightarrow \ D_f^*$

(8.61)                              $(A, \psi) \ \mapsto \ (y, x)$

*with $\tilde{D}_f = \{(A; \psi) : \ \tilde{c}_1 \varepsilon^{j/2} \leq A \leq \tilde{c}_2 L^{j/2}, \ \psi \in \mathbb{T}\}$, where $\tilde{c}_i$, $i = 1, 2$ are constants independent of $\varepsilon$, such that:*

   (1) $K_0 \circ \chi_f(A, \psi) = \mathcal{G}_f(A; \varepsilon)$.
   (2) $\|\chi_f\|_{\mathcal{C}^s(\tilde{D}_f)} \leq \frac{M_f}{\varepsilon^{sj}}$, $\quad \|\chi_f^{-1}\|_{\mathcal{C}^s(D_f^*)} \leq \frac{M_f}{\varepsilon^{sj}}$, $0 \leq s \leq r - 2m - 2$.
   (3) $\|\mathcal{G}_f\|_{\mathcal{C}^3(\tilde{D}_f)} \leq \frac{M_f}{\varepsilon^{j/2}}$ and $\left| \mathcal{G}_f''(A; \varepsilon) \right| \geq M_f$.

*where $M_f$ is a constant independent of $\varepsilon$.*

PROOF. In $D_f^*$ we consider the curves $E = K_0(y, x; \varepsilon)$, with $E \geq c_1 \varepsilon^j$. Then the action variable is given by the well known formula:

$$A = \frac{S(E)}{2\pi} = \frac{S(E; \varepsilon)}{2\pi} = \frac{1}{2\pi} \oint_{K_0^{-1}(E)} y \, dx,$$

and $\psi$ is the conjugate angle. The new Hamiltonian will be $\mathcal{G}_f(A; \varepsilon) = S^{-1}(A; \varepsilon)$.

To avoid an unpleasant typography, from now on in this section, we will omit the dependence on $\varepsilon$ of many of the functions which appear during the proof.

Following standard practice in mechanics we find it useful to use variables $(x, E)$ rather than $(y, x)$. (Note that we are interested in the curves $E = K_0(y, x; \varepsilon)$.) We denote by $\Gamma$ the map that to $(y, x)$ associates $(x, E)$. The mapping $\Gamma$ satisfies $\Gamma(D_f^*) \subset \mathcal{J}_f$, where

$$(8.62) \qquad \mathcal{J}_f = \{(x, E), \; x \in \mathbb{T}, \; c_1 \varepsilon^j \le E \le c_2 L\}.$$

We also note that the mapping $\Gamma$ is locally invertible.

Using part (1) of Lemma 8.34 with $\delta = \varepsilon^j$, and $n = r - 2m - 1$, we define in $D_f^*$:

$$E = K_0(y, x; \varepsilon),$$
$$\tau(x, E) = \int_0^x \frac{1}{\frac{\partial K_0}{\partial y}(\mathcal{Y}_\pm(u, E), u; \varepsilon)} \, du.$$

where $E$ is the energy of the orbit and $\tau$ is the time along the orbit of energy $E$ (we have chosen the origin of time at $x = 0$). With this choice, we have that $T(E) = \tau(2\pi, E)$ is the period of the periodic orbit, and the action-angle variables are:

$$(8.63) \qquad
\begin{aligned}
A &= \frac{S(E)}{2\pi} = \frac{1}{2\pi} \int_0^{2\pi} \mathcal{Y}_\pm(x, E) \, dx, \\
\psi &= \frac{2\pi}{T(E)} \tau(x, E),
\end{aligned}$$

where $S'(E) = T(E)$.

The region $D_f^*$ has two connected components. We will do all the details when $y > 0$. In the other region, the calculation is identical modulo adding some extra signs that disappear when we estimate the absolute values.

First, using implicit derivatives in equation (8.54) and formula (8.56), we have:

$$(8.64) \qquad
\begin{aligned}
\tau(x, E) &= \int_0^x \frac{1}{\frac{\partial K_0}{\partial y}(\mathcal{Y}_+(u, E), u; \varepsilon)} \, du = \int_0^x \frac{\partial \mathcal{Y}_+}{\partial E}(u, E) \, du \\
&= \frac{1}{\sqrt{2}} \int_0^x \frac{1 + \varepsilon b}{\sqrt{E - \varepsilon^j U(u; \varepsilon)}} \, du + \varepsilon P_1(x, E),
\end{aligned}$$

where the function $P_1(x, E)$ is given by $P_1(x, E) = \int_0^x \frac{\partial}{\partial E} \tilde{\mathcal{Y}}_+(\ell(u, E)) \, du$. Taking into account that in $D_f^*$, and so in $\mathcal{J}_f$ (see (8.62)), one has:

$$(8.65) \qquad c_1 \varepsilon^j \le E \le E - \varepsilon^j U(x; \varepsilon) \le E + c\varepsilon^j \le \text{cte.} \, E,$$

bound (8.57) give us that

$$(8.66) \qquad |P_1|_{\mathcal{C}^0(\mathcal{J}_f)} \le \text{cte.}, \quad |P_1|_{\mathcal{C}^s(\mathcal{J}_f)} \le \frac{\text{cte.}}{\varepsilon^{j(s+1/2)}}, \quad 1 \le s \le r - 2m - 2.$$

Differentiating (8.64) under the integral sign, and using (8.65) and (8.66), we obtain upper bounds for $\tau(x, E)$:

$$(8.67) \qquad |\tau|_{\mathcal{C}^s(\mathcal{J}_f)} \le \frac{\text{cte.}}{\varepsilon^{j(s+1/2)}}, \quad 0 \le s \le r - 2m - 2.$$

On the other hand, using that $\tau(2\pi, E) = S'(E)$, one obtains:

$$(8.68) \qquad |S|_{\mathcal{C}^0(\mathcal{J}_f)} \leq \text{cte.}, \quad |S|_{\mathcal{C}^s(\mathcal{J}_f)} \leq \frac{\text{cte.}}{\varepsilon^{j(s-1/2)}}, \quad 1 \leq s \leq r - 2m - 2.$$

The next task is to obtain lower bounds for the first and second derivatives of $S$. We start by observing that the main term of $\tau(x, E)$ appears explicitly in (8.64), and we have developed in (8.66) bounds for the remainder. Using (8.65) we can obtain easily lower bounds for the main term.

$$(8.69) \qquad\qquad\qquad |S'(E)| = |T(E)| \geq \frac{\text{cte.}}{E^{1/2}},$$

$$(8.70) \qquad\qquad\qquad\qquad |S''(E)| \geq \frac{\text{cte.}}{E^{3/2}}.$$

Once we have estimated the function $T(E)$, we can bound the conjugate angle $\psi$ given in (8.63), where $\tau(x, E)$ is given by (8.64). First, using (8.68), (8.69) and Faa-di-Bruno formula for the derivative of the composition of functions, we obtain

$$\left| \frac{\partial^s}{\partial E^s} \left( \frac{1}{T(E)} \right) \right| \leq \frac{\text{cte.}}{E^{s-1/2}} \leq \frac{\text{cte.}}{\varepsilon^{j(s-1/2)}},$$

and then the Leibniz rule gives

$$\left| \frac{1}{T} P_1 \right|_{\mathcal{C}^0(\mathcal{J}_f)} \leq \text{cte.}, \quad \left| \frac{1}{T} P_1 \right|_{\mathcal{C}^s(\mathcal{J}_f)} \leq \frac{\text{cte.}}{\varepsilon^{j(s-1)}}, \quad 1 \leq s \leq r - 2m - 2.$$

This gives, using again (8.65):

$$(8.71) \qquad\qquad \left| \frac{1}{T}\tau \right|_{\mathcal{C}^s(\mathcal{J}_f)} \leq \frac{\text{cte.}}{\varepsilon^{js}}, \quad 0 \leq s \leq r - 2m - 2.$$

Bound (8.71) together with (8.68) gives the upper bound claimed in item (2) of the Proposition for the $\mathcal{C}^s$ norm of the change of variables

$$\chi_f^{(-1)}(y, x) = \left( \frac{S(E)}{2\pi}, \frac{2\pi}{T(E)}\tau(x, E) \right),$$

with $E = K_0(y, x; \varepsilon)$, for $0 \leq s \leq r - 2m - 2$ in $D_f^*$.

Since $\det D\chi_f^{(-1)}(y, x) = 1$, the $\mathcal{C}^s$ norm of $\chi_f$ satisfies the same bounds.

Moreover, taking into account that $\mathcal{G}_f = S^{-1}$, the lower bound for the second derivatives of $\mathcal{G}_f$ and the upper bound for the third derivative of $\mathcal{G}_f$ follow from (8.69-8.70).                                                    $\square$

Now that once we have expressed $K_0$, given in (8.41), in action-angle variables in $D_f^*$, the proof of Theorem 8.30 will consist in applying Theorem 8.19 to the time-$2\pi k_0$ map of the full Hamiltonian (8.36), to show that it has primary invariant tori. Going back to the original variables $(y, x, s)$, one obtains the result claimed in Theorem 8.30. We proceed to give the details.

PROOF OF PART 1) OF THEOREM 8.30. The proof will consist in applying Theorem 8.19 to $F$, the time-$2\pi k_0$ map of the Hamiltonian

$$(8.72) \qquad \tilde{K}(A, \psi, s; \varepsilon) = \mathcal{G}_f(A; \varepsilon) + \varepsilon^{m+1} \tilde{S}(A, \psi, s; \varepsilon),$$

where $\tilde{K}(A, \psi, s; \varepsilon) = K \circ \chi_f(A, \psi, s; \varepsilon)$, and $\tilde{S}(A, \psi, s; \varepsilon) = S \circ \chi_f(A, \psi, s; \varepsilon)$.

Since $\tilde{K}$, $\mathcal{G}$, $\tilde{S}$ are $\mathcal{C}^{r-2m-2}$, and we have assumed in Theorem 8.30 that $r - 2m - 2 \geq 6$, we have that $\tilde{K}$, $\mathcal{G}$, $\tilde{S}$ are $\mathcal{C}^6$.

We denote by $F_0$ the time-$2\pi k_0$ map of Hamiltonian $\mathcal{G}_f(A; \varepsilon)$ and we have that $F$ and $F_0$ are $\mathcal{C}^5$ and we can bound

$$||F - F_0||_{\mathcal{C}^5} \leq \text{cte.} \, \varepsilon^{m+1} ||\tilde{S}||_{\mathcal{C}^6} \leq \text{cte.} \, \varepsilon^{m+1-6j}.$$

Since $\mathcal{G}_f$ depends only on $A$, $F_0$ is an integrable map of the form $(A, \Psi) \mapsto (A, \Psi + \Delta(A))$.

Furthermore, by item (3) of Proposition 8.35, we have

$$\frac{d}{dA} \Delta(A) = \frac{\partial^2}{\partial A^2} \mathcal{G}_f(A; \varepsilon) \geq M_f.$$

Hence, the mapping $F_0$ will be a twist mapping, and we can apply Theorem 8.19 with $\delta = \varepsilon^{m+1-6j}$, and we obtain, if $m > 6j - 1$:

(1) There exists a set of values $A_i$, such that the Hamiltonian $K \circ \chi_f$ has invariant tori given by

$$A = A_i + \mathcal{A}_i(\psi, s; \varepsilon),$$

where $\mathcal{A}_i$ are $\mathcal{C}^{r-2m-5-\eta}$ functions, for any $\eta > 0$, and $||\mathcal{A}_i||_{\mathcal{C}^2} \leq \text{cte.} \, \varepsilon^{(m+1-6j)/2}$.

(2) The motion on these tori is $\mathcal{C}^1$-conjugate to a rigid translation of frequencies $(\omega(A_i), 1)$, where $\omega(A_i)$ is a Diophantine number of constant type and Markov constant $K \varepsilon^{\frac{m+1-6j}{2}}$, as stated in Definition 8.16.

(3) The union of neighborhoods of size $\varepsilon^{(m+1-6j)/2}$ of these tori cover all the region $\tilde{D}_f$.

Going back to the original variables $(y, x, s) = (\chi_f(A, \psi), s)$ , and using that $K_0(y, x; \varepsilon) = E = \mathcal{G}(A; \varepsilon)$, and that $||\mathcal{G}_f||_{\mathcal{C}^3} \leq \frac{\text{cte.}}{\varepsilon^{j/2}}$, and $||\chi_f^{-1}||_{\mathcal{C}^2} \leq \frac{\text{cte.}}{\varepsilon^{2j}}$, one obtains the desired result if $m + 1 - 11j > 0$, calling $E_i = \mathcal{G}(A_i; \varepsilon)$. $\qquad \square$

Now we turn to estimate the action-angle variables in the regions $D_{o,in}$. We use that, in $D_{o,in}$, the coordinate $y$ is of size $\varepsilon^{j/2}$ and then, it is natural to perform the scaling $y = \varepsilon^{j/2} Y$, which leads directly to the following Lemma. This transformation is not symplectic, but conformally symplectic so that the equations are still in Hamiltonian form.

LEMMA 8.36. *The scaling $y = \varepsilon^{j/2} Y$ transforms the Hamiltonian system of Hamiltonian $K(y, x, s; \varepsilon)$ given in (8.36) into a Hamiltonian system of*

$\mathcal{C}^{r-2m-2}$ *Hamiltonian*

$$
\begin{aligned}
\mathcal{K}(Y,x,s;\varepsilon) &= \varepsilon^{j/2}(h_{\text{int}}(Y;\varepsilon^{j/2}) + U(x;\varepsilon)) + \varepsilon^{m+1-j/2}S_1(Y,x,s;\varepsilon^{j/2}) \\
(8.73) \qquad &= \varepsilon^{j/2}\mathcal{K}_{\text{int}}(Y,x;\varepsilon^{j/2}) + \varepsilon^{m+1-j/2}S_1(Y,x,s;\varepsilon^{j/2})
\end{aligned}
$$

*with*

$$
\begin{aligned}
h_{\text{int}}(Y;\varepsilon^{j/2}) &= \frac{Y^2}{2}\hat{h}(\varepsilon^{j/2}Y;\varepsilon), \\
S_1(Y,x,s;\varepsilon^{j/2}) &= S(\varepsilon^{j/2}Y,x,s;\varepsilon),
\end{aligned}
$$

*where $\hat{h}(y,\varepsilon)$ is given in (8.39) and, consequently, $\mathcal{K}_{\text{int}}$ is a $\mathcal{C}^{r-2m-1}$ function.*

*The scaling transforms the domains $D_{o,in}$ given in (8.44), (8.45) into*

$$
\begin{aligned}
\mathcal{D}_o &= \{(Y,x,s) : \mathcal{K}_{\text{int}}(Y,x;\varepsilon^{j/2}) = F/\varepsilon^j, \; c_3\varepsilon^\alpha \le F \le c_1\varepsilon^j\} \\
(8.74) \qquad &= \{(Y,x,s) : \mathcal{K}_{\text{int}}(Y,x;\varepsilon^{j/2}) = e, \; c_3\varepsilon^{\alpha-j} \le e \le c_1\}, \\
\mathcal{D}_{in} &= \{(Y,x,s) : \mathcal{K}_{\text{int}}(Y,x;\varepsilon^{j/2}) = G/\varepsilon^j, \; -c_4\varepsilon^j \le G \le -c_3\varepsilon^\alpha\} \\
(8.75) \qquad &= \{(Y,x,s) : \mathcal{K}_{\text{int}}(Y,x;\varepsilon^{j/2}) = e, \; -c_4 \le e \le -c_3\varepsilon^{\alpha-j}\}.
\end{aligned}
$$

In the above formulas $(x,s) \in (\mathbb{R}/(2\pi k_0\mathbb{Z}))^2$. We define the action angle variables in

$$
(8.76) \qquad \mathcal{D}_o^* = \{(Y,x) \in \mathbb{R} \times \mathbb{T}, \; \exists l \in \{0,1,\cdots,k_0\}, \; (Y,x+2\pi l,s) \in \mathcal{D}_o\}
$$

by formulas (8.63) as in $\mathcal{D}_f^*$. The only change is that, in $\mathcal{D}_o^*$, instead of (8.65) we have:

$$
(8.77) \qquad 0 \le c_3\varepsilon^{\alpha-j} \le e \le e - U(x;\varepsilon) \le e + c \le c_1 + c.
$$

To bound accurately the action-angle variables for Hamiltonian $\mathcal{K}_{\text{int}}(Y,x;\varepsilon^{j/2})$ in $\mathcal{D}_o^*$ we use the following:

LEMMA 8.37. *There exist $A_1$, $A_2$, independent of $\varepsilon$, $e$, such that if $0 < c_3\varepsilon^{\alpha-j} \le e \le c_1$ then, for any $x \in [0,2\pi]$*

(1)

$$
A_1 \log(1/e) \le \int_0^x \frac{du}{\sqrt{e - U(u;\varepsilon)}} \le A_2 \log(1/e).
$$

(2) *If $n \ge 2$*

$$
\frac{A_1}{e^{(n-1)/2}} \le \int_0^x \frac{du}{(e - U(u;\varepsilon))^{n/2}} \le \frac{A_2}{e^{(n-1)/2}}.
$$

PROOF. As we have by (8.38) that $U(x;\varepsilon) \simeq -ax^2 + O(x^3)$, as $x \to 0$, we know that, there exists some $\rho > 0$, independent of $\varepsilon$, and $a_i = a_i(\rho)$ such that

$$
(8.78) \qquad a_1 x^2 \le -U(x;\varepsilon) \le a_2 x^2 \quad \forall x \in [-\rho,\rho]
$$

Then, we compute

$$
\int_0^x \frac{1}{(e - U(u;\varepsilon))^{n/2}} du
$$

(8.79)
$$
= \int_0^\rho \frac{1}{(e - U(u;\varepsilon))^{n/2}} du + \int_\rho^{x-\rho} \frac{1}{(e - U(u;\varepsilon))^{n/2}} du
$$

$$
+ \int_{x-\rho}^x \frac{1}{(e - U(u;\varepsilon))^{n/2}} du.
$$

When $x < \rho$, we will not divide the integral, but we will analyze it as we analyze the first term in (8.79).

We will compute carefully the first integral in the right hand side of (8.79). Similar estimates are valid for the third integral when $x$ is close to $2\pi$, and the second is a bounded function even for $0 \le e \le c_1$.

To estimate the first integral in the right hand side of (8.79), we use (8.78), and the formula

$$
\int_0^\rho \frac{du}{(e + a_i u^2)^{n/2}} = \frac{1}{\sqrt{a_i} e^{(n-1)/2}} \int_0^{\frac{\sqrt{a_i}}{\sqrt{e}}\rho} \frac{dt}{\sqrt{(1 + t^2)^n}}.
$$

If $n = 1$, this integral diverges as $e \to 0$, but integrating explicitly we get

$$
\frac{1}{\sqrt{a_i}} \int_0^{\frac{\sqrt{a_i}}{\sqrt{e}}\rho} \frac{1}{(1 + t^2)^{1/2}} = \frac{1}{\sqrt{a_i}} \log\left( \frac{\sqrt{a_i}}{\sqrt{e}}\rho + \sqrt{1 + \frac{a_i}{e}\rho^2} \right).
$$

If $n \ge 2$, the integral is convergent for $e = 0$. This gives us the bounds of the Lemma 8.37. $\qquad\square$

The next Proposition 8.38 is devoted to expressing the integrable Hamiltonian $\mathcal{K}_{\text{int}}(Y, x; \varepsilon^{j/2})$ into action-angle variables $(A, \psi)$ in the regions $\mathcal{D}^*_{o,in}$, where $\mathcal{D}^*_0$ is defined in (8.76) and $\mathcal{D}^*_{in}$ is defined analogously.

PROPOSITION 8.38. *Consider the $\mathcal{C}^{r-2m-1}$ Hamiltonian $\varepsilon^{j/2}\mathcal{K}_{\text{int}}(Y, x; \varepsilon^{j/2})$ in the regions $\mathcal{D}^*_o$, $\mathcal{D}^*_{in}$. Then, for $\upsilon = o, in$, there exist $\mathcal{C}^{r-2m-2}$ changes of variables*

$$
\chi_\upsilon : \tilde{\mathcal{D}}_\upsilon \;\to\; \mathcal{D}^*_\upsilon
$$
$$
(A, \psi) \;\mapsto\; (Y, x)
$$

*where $\tilde{\mathcal{D}}_\upsilon = \{(A, \psi) : \tilde{c}^1_\upsilon \le A \le \tilde{c}^2_\upsilon\}$, and $\tilde{c}^l_\upsilon$, $l = 1, 2$, are suitable constants independent of $\varepsilon$, such that:*

(1) $\varepsilon^{j/2}\mathcal{K}_{\text{int}} \circ \chi_\upsilon(A, \psi) = \varepsilon^{j/2}\mathcal{G}_\upsilon(A; \varepsilon^{j/2})$.
(2) *For $0 \le s \le r - 2m - 2$ we have that:*

$$
\|\chi_\upsilon\|_{\mathcal{C}^s(\tilde{\mathcal{D}}_\upsilon)} \le \frac{M_\upsilon}{\varepsilon^{s(\alpha-j)}}, \quad \|\chi_\upsilon^{-1}\|_{\mathcal{C}^s(\mathcal{D}^*_\upsilon)} \le \frac{M_\upsilon}{\varepsilon^{s(\alpha-j)}}.
$$

(3) $\|\mathcal{G}_\upsilon\|_{\mathcal{C}^3(\tilde{\mathcal{D}}_\upsilon)} \le \frac{M_\upsilon}{\varepsilon^{2(\alpha-j)}}$, $|\mathcal{G}''_\upsilon(A, \varepsilon^{j/2})| \ge M_\upsilon$.

*where $M_\upsilon$ is a constant independent of $\varepsilon$.*

PROOF. As in Proposition 8.35, we consider the curves $e = \mathcal{K}_{\text{int}}(y, x; \varepsilon^{j/2})$, and using Lemma 8.34 with $\delta = 1$, we define in $D_o^*$:

$$
\begin{aligned}
e &= \mathcal{K}_{\text{int}}(y, x; \varepsilon^{j/2}), \\
\tau(x, e) &= \int_0^x \frac{1}{\frac{\partial \mathcal{K}_{\text{int}}}{\partial y}(\mathcal{Y}_\pm(u, e), u; \varepsilon^{j/2})} du.
\end{aligned}
$$

Then, $T(e) = \tau(2\pi, e)$ is the period of the periodic orbit, and the action-angle variables are given by:

$$
\begin{aligned}
A &= \frac{S(e)}{2\pi} = \frac{1}{2\pi} \int_0^{2\pi} \mathcal{Y}_\pm(x, e) dx, \\
\psi &= \frac{2\pi}{T(e)} \tau(x, e),
\end{aligned}
$$

where $S'(e) = T(e)$.

The region $D_o^*$ has two connected components, we will give full details for the component where $y > 0$. The other one is identical.

Remember that, when $(Y, s) \in \mathcal{D}_o^*$, the variables $(x, e)$ are in $\mathcal{J}_o$:

$$
(8.80) \qquad \mathcal{J}_o = \{(x, e), \ x \in \mathbb{T}, \ c_3 \varepsilon^{\alpha-j} \le e \le c_1\}
$$

First, using equation (8.56), we have, as in (8.64):

$$
(8.81) \qquad \tau(x, e) = \frac{1}{\sqrt{2}} \int_0^x \frac{1 + \varepsilon b}{\sqrt{e - U(u; \varepsilon)}} du + \varepsilon P_1(x, e),
$$

where $P_1(x, e) = \int_0^x \frac{\partial}{\partial e} \tilde{\mathcal{Y}}_+(\ell(u, e)) du$. Bounds (8.57) and (8.77) give, in $D_o$, and then, in $\mathcal{J}_o$:

$$
|P_1|_{\mathcal{C}^0(\mathcal{J}_o)} \le \text{cte.}, \ |P_1|_{\mathcal{C}^s(\mathcal{J}_o)} \le \frac{\text{cte.}}{\varepsilon^{(\alpha-j)(s-1/2)}}, \ 1 \le s \le r - 2m - 2.
$$

From these inequalities and Lemma 8.37, we obtain upper bounds for $\tau(x, e)$:

$$
(8.82) \quad |\tau|_{\mathcal{C}^0(\mathcal{J}_o)} \le \text{cte.} \log \frac{1}{\varepsilon^{\alpha-j}}, \ |\tau|_{\mathcal{C}^s(\mathcal{J}_o)} \le \frac{1}{\varepsilon^{(\alpha-j)s}}, \ 1 \le s \le r - 2m - 2.
$$

On the other hand, from $\tau(2\pi, e) = S'(e)$, we obtain:

$$
(8.83) \qquad |S|_{\mathcal{C}^0(\mathcal{J}_o)} \le \text{cte.}, \ |S|_{\mathcal{C}^s(\mathcal{J}_o)} \le \frac{\text{cte.}}{\varepsilon^{(\alpha-j)(s-1)}}, \ 1 \le s \le r - 2m - 2,
$$

and, using again Lemma 8.37, we also obtain lower bounds for the first and second derivatives of $S$:

$$
(8.84) \qquad \qquad |S'(e)| = |T(e)| \ge \text{cte.} \log(1/e),
$$

$$
(8.85) \qquad \qquad |S''(e)| \ge \frac{\text{cte.}}{e}.
$$

To bound $\psi = \frac{2\pi}{T(e)} \tau(x, e)$ we use (8.84), (8.83) and Faà-di-Bruno formulas for the composition of functions derivatives, obtaining:

$$
\left| \frac{\partial^s}{\partial e^s} \frac{1}{T(e)} \right| \le \frac{\text{cte.}}{\varepsilon^{(\alpha-j)s}}, \ 0 \le s \le r - 2m - 2,
$$

then, the Leibniz rule gives:

$$\left|\frac{1}{T}P_1\right|_{\mathcal{C}^0(\mathcal{J}_o)} \leq \text{cte.} \,, \quad \left|\frac{1}{T}P_1\right|_{\mathcal{C}^s(\mathcal{J}_o)} \leq \frac{\text{cte.}}{\varepsilon^{(\alpha-j)(s-1/2)}}, \quad 1 \leq s \leq r - 2m - 2.$$

This gives, using again (8.77),

$$\left|\frac{1}{T}\tau\right|_{\mathcal{C}^s(\mathcal{J}_o)} \leq \frac{\text{cte.}}{\varepsilon^{(\alpha-j)s}}, \quad 0 \leq s \leq r - 2m - 2.$$

This bound together with (8.83) establishes the bound claimed in statement (2) of the Proposition 8.38 for the $\mathcal{C}^s$ norm of the change of variables

$$\chi_v^{(-1)}(y,x) = \left(\frac{S(e)}{2\pi}, \frac{2\pi}{T(e)}\tau(x,e)\right)$$

with $e = K_{\text{int}}(Y,x;\varepsilon^{j/2})$, $0 \leq s \leq r - 2m - 2$.

Since $\det D\chi_v^{(-1)}(Y,x) = 1$, the $\mathcal{C}^s$ norm of $\chi$ satisfies the same estimates.

Moreover, taking into account that $\mathcal{G}_f = S^{-1}$, the lower bound for the second derivatives of $\mathcal{G}_f$ and the upper bound for the third derivative of $\mathcal{G}_f$ follow from (8.84), (8.83), (8.85).

This finishes the proof of Proposition 8.38 for the region $\mathcal{D}_o^*$.

In the region $\mathcal{D}_{in}^*$ the action variable is defined as

$$(8.86) \qquad A = \frac{S(e)}{2\pi} = \frac{1}{2\pi}\oint_{K_{\text{int}}^{-1}(e)} Y\, dx,$$

and $\psi$ is the conjugate angle.

The main complication to obtain the bounds of the change of variables is that the curves $K_{\text{int}}^{(-1)}(e)$ when $e \leq 0$, cannot be parameterized as graphs of functions $Y = \mathcal{Y}(x,e)$ so that the computation of (8.86) will need to be divided in different pieces.

Choosing the origin of time in the section

$$\Sigma = \{(Y,x) : Y > 0, x = \pi\},$$

we define $t(Y,x)$ as the time that the trajectory starting in $(Y,x)$ takes to arrive to the section $\Sigma$.

To understand the regularity properties of the function $t(Y,x)$ we find it convenient to perform different arguments in different regions of the $(Y,x)$ plane.

Let $\rho > 0$ be the number that appears in Lemma 8.34, part (2).

If $(Y,x) \in \mathcal{D}_{in,\rho}^* \equiv \{(Y,x) \in K_{\text{int}}^{(-1)}(e), \; Y > 0, \; \rho < x < 2\pi - \rho\}$, the function $t(Y,x)$ is of class $\mathcal{C}^{r-2m-2}$ and its $\mathcal{C}^{r-2m-2}$ norm is bounded independently of $e, \varepsilon$.

Outside of $\mathcal{D}_{in,\rho}^*$ the function $t(Y,x)$ is more complicated to analyze since the orbits starting outside of $\mathcal{D}_{in,\rho}^*$ can pass close to the critical point $(0,0)$ before reaching $\Sigma$, and the time $t(Y,x)$ goes to infinity when $e$ goes to zero.

We will explain the case when $Y > 0$ and $0 < x < \rho$, and the other cases are easy modifications of this one.

If $x < \rho$, and $Y > 0$, we compute the time as
$$t(Y,x) = t_1(Y,x) + t_2(Y,x),$$
where $t_1$ is the time to arrive at the section given by $x = \rho$ (we denote by $Y(\rho)$ the corresponding $Y$ coordinate: $\mathcal{K}_{int}(Y(\rho), x; \varepsilon^{j/2}) = e$), and $t_2$ is the time between the sections $x = \rho$ and $\Sigma$. Note that $t_2$ is $\mathcal{C}^{r-2m-2}$ with a bounded $\mathcal{C}^{r-2m-2}$ norm, so we only need to estimate $t_1$.

To bound $t_1$, we use Lemma 8.34 and the Hamiltonian equations obtaining that

$$
\begin{aligned}
t_1(Y,x) &= -\int_Y^{Y(\rho)} \frac{du}{\frac{\partial \mathcal{K}_{int}}{\partial x}(u, \mathcal{X}_+(u,e); \varepsilon^{j/2})} \\
&= -\int_Y^{Y(\rho)} \frac{\partial \mathcal{X}_+}{\partial e}(u,e)\,du \\
&= \frac{1}{\sqrt{a}} \int_Y^{Y(\rho)} \frac{1}{\sqrt{h_{int}(u; \varepsilon^{j/2}) - e}}\,du + \tilde{P}_1(Y,e).
\end{aligned}
$$

This formula is very similar to formula (8.81) and analogous arguments to the ones used there establish that

$$|t_1|_{\mathcal{C}^0(\mathcal{J}_i)} \le \text{cte. } \log(1/e)), \quad |t_1|_{\mathcal{C}^s(\mathcal{J}_i)} \le \frac{\text{cte.}}{\varepsilon^{(\alpha-j)s}}, \quad 1 \le s \le r - 2m - 2.$$

where $\mathcal{J}_i$ is defined analogously to $\mathcal{J}_o$ in (8.80) to be the domain of the variables $(x,e)$.

Analogous arguments can be used in the cases when $Y < 0$, or $x < \rho$, obtaining exactly the same kind of bounds in all $D_{in}^*$. Once we have defined the function $t(Y,x)$ in $D_{in}^*$, we have that the action-angle variables are given by

$$
\begin{aligned}
A &= \frac{S(e)}{2\pi} \\
\psi &= 2\pi \frac{t(Y,x)}{T(e)},
\end{aligned}
$$

where $T(e) = S'(e)$. Proceeding as we did in $D_o^*$, we obtain the same kind of bounds in these two regions. This finishes the proof of Proposition 8.38.  $\square$

Applying again Theorem 8.19 to the time-$2\pi k_0$ map of Hamiltonian (8.36) after it has been scaled and written in action-angle variables and going back to the original variables $(y, x, s)$ we obtain the existence of primary tori in $D_o^*$ and secondary tori in $D_{in}^*$.

PROOF OF PARTS 2) AND 3) OF THEOREM 8.30. We will prove the results in both regions $D_o^*$, $D_{in}^*$ at the same time by composing the Hamiltonian $\mathcal{K}$ with the two different $\mathcal{C}^{r-2m-2}$ changes $\chi_v$, $v = o, in$.

Again, the proof will consist in applying Theorem 8.19 to $F$, the time-$2\pi k_0$ map of the Hamiltonian

$$(8.87) \qquad \tilde{\mathcal{K}}_v(A, \psi, s; \varepsilon) = \varepsilon^{j/2} \mathcal{G}_v(A; \varepsilon^{j/2}) + \varepsilon^{m+1-j/2} \tilde{S}_v(A, \psi, s; \varepsilon^{j/2}),$$

where $\tilde{K}_v = \mathcal{K} \circ \chi_v$ and $\tilde{S}_v = S_1 \circ \chi_v$.

Since $\tilde{K}_v$, $\mathcal{G}_v$, $\tilde{S}_v$ are $\mathcal{C}^{r-2m-2}$, and we have assumed in Theorem 8.30 that $r - 2m - 2 \geq 6$, we have that $\tilde{K}_v$, $\mathcal{G}_v$, $\tilde{S}_v$ are $\mathcal{C}^6$.

We denote by $F_0$ the time-$2\pi k_0$ map of Hamiltonian $\varepsilon^{j/2}\mathcal{G}_v(A; \varepsilon^{j/2})$. We have that $F$ and $F_0$ are $\mathcal{C}^5$ and using standard results on dependence of solutions of ode's on parameters

$$||F - F_0||_{\mathcal{C}^5} \leq \text{cte.}\, \varepsilon^{m+1-j/2}||\tilde{S}_v||_{\mathcal{C}^6} \leq \text{cte.}\, \varepsilon^{m+1-j/2-6(\alpha-j)}.$$

Since $\mathcal{G}_f$ depends only on $A$, $F_0$ is an integrable map of the form $(A, \Psi) \rightarrow (A, \Psi + \Delta(A))$.

Furthermore, by item (3) of Proposition 8.38, we have

$$\frac{d}{dA}\Delta(A) = \varepsilon^{j/2}\frac{\partial^2}{\partial A^2}\mathcal{G}_v(A; \varepsilon^{j/2}) \geq M_v\varepsilon^{j/2}.$$

Hence, the mapping $F_0$ will be a twist mapping, and we can apply Theorem 8.19 with $\delta = \varepsilon^{m+1-j/2-6(\alpha-j)}$, and we obtain, for $m > 6(\alpha-j)+j/2-1$:

(1) There exist a set of values $A_l$, such that he Hamiltonian $\tilde{\mathcal{K}}$ has invariant tori given by

$$A = A_l + \mathcal{A}_l(\psi, s; \varepsilon^{j/2}),$$

where $\mathcal{A}_l$ are $\mathcal{C}^{r-2m-5-\eta}$ functions, for any $\eta > 0$, and $||\mathcal{A}_l||_{\mathcal{C}^2} \leq$ cte. $\varepsilon^{\frac{m+1-6(\alpha-j)-3j/2}{2}}$.

(2) The motion on these tori is $\mathcal{C}^1$-conjugate to a rigid translation of frequencies $(\omega(A_l), 1)$, where $\omega(A_l)$ is a Diophantine number of constant type and Markov constant $K\varepsilon^{\frac{m+1-6(\alpha-j)-j/2}{2}}$, as stated in Definition 8.16.

(3) The union of neighborhoods of size cte. $\varepsilon^{\frac{m+1-6(\alpha-j)-3j/2}{2}}$ of these tori cover all the regions $\tilde{D}_v$.

Going back to the variables $(Y, x, s)$, and using that

$$||\mathcal{G}_v||_{\mathcal{C}^3} \leq \text{cte.}\, \varepsilon^{-2(\alpha-j)}$$

$$||\chi_v^{-1}||_{\mathcal{C}^2(\mathcal{D}_v^*)} \leq \text{cte.}\, \varepsilon^{-2(\alpha-j)},$$

we obtain, for $m > 14(\alpha - j) + 3j/2 - 1$, that the tori are given by

$$\mathcal{K}_{\text{int}}(Y, x; \varepsilon^{j/2}) = e_l + O_{\mathcal{C}^2}\left(\varepsilon^{\frac{m+1-3j/2-14(\alpha-j)}{2}}\right).$$

Multiplying this equality by $\varepsilon^j$ and performing the scaling $y = \varepsilon^{j/2}Y$ we obtain that the tori are given by

$$K_0(y, x; \varepsilon) = \varepsilon^j e_l + O_{\mathcal{C}^2}\left(\varepsilon^{\frac{m+1+j/2-14(\alpha-j)}{2}}\right).$$

Now, calling $F_l = \varepsilon^j e_l$, for $v = o$, and $G_l = \varepsilon^j e_l$, for $v = i$, we have the results claimed in Theorem 8.30. $\qquad\square$

Before proving Corollary 8.31, we analyze the equation

$$(8.88) \qquad K_0(y, x; \varepsilon) = E + \nu g(y, x, s, E; \varepsilon).$$

LEMMA 8.39. *Let us consider equation* (8.88) *where* $K_0(y, x; \varepsilon)$ *is given in* (8.41) *and* (8.39), *and* $g(y, x, s, E; \varepsilon)$ *is at least of class* $\mathcal{C}^2$, *with*

$$(8.89) \qquad \|g\|_{\mathcal{C}^2} \leq \text{cte.}$$

*for* $|y| \leq c_2 L$ *and*

   a) $c_3 \varepsilon^\alpha \leq E \leq c_2 \bar{L}$ *and* $(x, s) \subset \mathbb{T}^2$

*or*

   b) $-c_4 \varepsilon^j \leq E \leq 0$, $(x, s) \in \mathbb{T}^2$, *and* $x \in [\rho, 2\pi - \rho]\}$, *where* $0 < \rho < \pi$
   *is any number independent of* $\varepsilon, \nu$.

*Then, for* $\alpha > j$, $j = 1, 2$, *and for* $\varepsilon$ *small enough, there exists some* $\nu_0$, *independent of* $\varepsilon$, *such that if* $\nu \leq \nu_0 \varepsilon^\alpha$, *equation* (8.88) *defines a function* $y = f^\pm(x, s, E; \varepsilon, \nu)$ *of class* $\mathcal{C}^2$ *in the domains a) or b), such that:*

   (1) $f^\pm(x, s, E; \varepsilon, 0) = \mathcal{Y}_\pm(x, E)$, *where* $\mathcal{Y}_\pm(x, E)$ *are the functions* (8.56)
       *introduced in Lemma 8.34, with* $\delta = \varepsilon^j$.
   (2) *If we denote by* $Df = \frac{\partial f}{\partial x}, \frac{\partial f}{\partial s}$, *we have, for* $f = f^\pm$:

$$(8.90) \qquad |f| \leq \text{cte.}, \quad |Df| \leq \varepsilon^{j/2}, \quad |D^2 f| \leq \varepsilon^{j/2},$$

   *and:*

$$(8.91) \qquad \left|\frac{\partial f}{\partial E}\right| \leq \varepsilon^{-j/2}, \quad \left|\frac{\partial Df}{\partial E}\right| \leq \varepsilon^{-j/2}.$$

$$(8.92) \qquad |f - \mathcal{Y}| \leq \nu \varepsilon^{-j/2}, \quad |D(f - \mathcal{Y})| \leq \nu \varepsilon^{-j/2},$$

PROOF. By part (1) of Lemma 8.34 with $\delta = \varepsilon^j$, for $y > 0$, equation (8.88) is equivalent to the equation:

$$(8.93) \qquad M(y, x, s, E; \nu) \equiv y - \mathcal{Y}_+(x, t) = 0, \quad t = E + \nu g(y, x, s, E; \varepsilon)$$

where $\mathcal{Y}_+(x, t)$, is the function (8.56).

Differentiating (8.56), and using (8.57), for $\text{cte.}\, \varepsilon^\alpha \leq E - \text{cte.}\, |\nu| \leq |t| \leq E + \text{cte.}\, |\nu| \leq \text{cte.}\, \bar{L}$, one has in a) the following bounds for $\mathcal{Y}_+(x, E)$

$$(8.94) \qquad \begin{aligned} &\left|\frac{\partial \mathcal{Y}_+}{\partial x}\right| \leq \text{cte.}\, \varepsilon^{j/2}, \quad \left|\frac{\partial \mathcal{Y}_+}{\partial t}\right| \leq \text{cte.}\, \varepsilon^{-j/2} \\ &\left|\frac{\partial^2 \mathcal{Y}_+}{\partial x^2}\right| \leq \text{cte.}\, \varepsilon^{j/2}, \quad \left|\frac{\partial^2 \mathcal{Y}_+}{\partial x, \partial t}\right| \leq \text{cte.}\, \varepsilon^{-j/2}, \quad \left|\frac{\partial^2 \mathcal{Y}_+}{\partial t^2}\right| \leq \varepsilon^{-3j/2}. \end{aligned}$$

Using (8.94) and (8.89), one has:

$$(8.95) \qquad \left|\frac{\partial M}{\partial y} - 1\right| \leq \text{cte.}\, \nu \varepsilon^{-j/2},$$

so that, the implicit function theorem applied to (8.93), gives the existence of $f^+(x, s, E; \varepsilon, \nu)$ if $\nu/\varepsilon^{j/2} \leq \nu_0$, for some $\nu_0$ small enough.

In order to bound the derivatives of $f^+$, we take implicit derivatives in equation (8.93). Then, taking into account (8.94), and using that $|\nu| < \nu_0 \varepsilon^\alpha < \nu_0 \varepsilon^j$, we see that:

$$\left| \frac{\partial M}{\partial \tau} \right| \leq \text{cte.}\, \varepsilon^{j/2}, \ \tau = x, s, \qquad \left| \frac{\partial M}{\partial \nu} \right| \leq \text{cte.}\, \varepsilon^{-j/2},$$

$$\left| \frac{\partial^2 M}{\partial \tau_1 \tau_2} \right| \leq \text{cte.}\, \varepsilon^{j/2}, \ \tau_1, \tau_2 = y, x, s, \qquad \left| \frac{\partial^2 M}{\partial \nu \partial \tau} \right| \leq \text{cte.}\, \varepsilon^{-j/2}, \ \tau = y, x, s.$$

These bounds and (8.95), give the desired bounds (8.90) and (8.92).

Moreover, using (8.94) we see that:

$$\left| \frac{\partial M}{\partial E} \right| \ \leq \ \text{cte.}\, \varepsilon^{-j/2},$$

$$\left| \frac{\partial^2 M}{\partial \tau \partial E} \right| \ \leq \ \text{cte.}\, \varepsilon^{-j/2}, \ \tau = y, x, s.$$

These inequalities and (8.95) give the desired bounds (8.91) in the domain a) for $y > 0$.

An analogous proof gives the bounds in the domain a) for $y < 0$.

For negative values of $E$—and consequently of $t$—, we note that $\mathcal{Y}(x, t)$ is the composition of a regular function with the function

$$\ell(x, t) = \sqrt{2}\sqrt{t - \varepsilon^j U(x, \varepsilon)}.$$

We note that in the domain b)—recall that $x$ is restricted to some interval $[\rho, 2\pi - \rho]$—the function $\ell(x, t)$, verifies bounds (8.94) and therefore so does $\mathcal{Y}_\pm$.

$\square$

PROOF OF COROLLARY 8.31. From now on, we consider $\alpha = j/2 + 3/2$, $j = 1, 2$ and $m \geq 26$. We apply Lemma 8.39 to the implicit equations (8.48), (8.50) and (8.51), of the invariant tori given by Theorem 8.30.

The equation (8.48), that gives implicitly the tori $\mathcal{T}_{E_i}^\pm$ in $\mathcal{D}_f$, is a particular case of equation (8.88) taking $E = E_i$ and $\nu = \varepsilon^{(m+1-11j)/2}$. Analogously for equation (8.50) which gives implicitly the tori $\mathcal{T}_{F_i}^\pm$ in $\mathcal{D}_o$, taking $E = F_i$ and $\nu = \varepsilon^{(m+1+j/2-14(\alpha-j))/2}$ and equation (8.51), which gives the tori $\mathcal{T}_{G_i}$ in $\mathcal{D}_{in}$, if we take $E = G_i$ and $\nu = \varepsilon^{(m+1+j/2-14(\alpha-j))/2}$.

For $m \geq 26$ and $\alpha = 3/2 + j/2$, the condition $|\nu| \leq \varepsilon^\alpha$ of Lemma 8.39 is verified in the three cases of the previous paragraph. The results of Lemma 8.39 give us the items (1), (2), (3), and (6) a), (6) b) of Corollary 8.31.

From 1.5, 2.5 and 3.5 of theorem 8.30 we obtain

$$|E_i - E_{i+1}| \ \leq \ \varepsilon^{\frac{m+1-7j}{2}},$$

$$|F_i - F_{i+1}| \ \leq \ \varepsilon^{\frac{m+1+j/2-10(\alpha-j)}{2}},$$

$$|G_i - G_{i+1}| \ \leq \ \varepsilon^{\frac{m+1+j/2-10(\alpha-j)}{2}},$$

Taking into account the definitions (8.43), (8.44), (8.45) of $D_f$, $D_o$, $D_{in}$, we get

$$|E_1 - F_{l_F}| \leq \varepsilon^{\frac{m+1-7j}{2}},$$

$$|F_1 - G_{l_G}| \leq \varepsilon^\alpha + \varepsilon^{\frac{m+1+j/2-10(\alpha-j)}{2}}.$$

Since $\alpha = j/2 + 3/2$ and $m \geq 26$, all these exponents are bigger than $\frac{3}{2} + \frac{j}{2}$. Using the bounds (8.91) of Lemma 8.39, and the inequalities above, we obtain the bounds of (6) c).                                                         $\square$

### 8.5.5. Existence of stable and unstable manifolds of periodic orbits.
We recall that in Section 8.5.3 we had shown that, in appropriate variables, the motion in the region $\mathcal{S}^{\mathcal{R}_j}$, $j = 1, 2$, is described by the $\mathcal{C}^{r-2m-2}$ Hamiltonian $K(y, x, s; \varepsilon)$ given in (8.36).

The main part of the Hamiltonian $K$ in (8.36) is the $\mathcal{C}^{r-2m-1}$ term $K_0(y, x; \varepsilon)$ in (8.41), which has a saddle at $(0, 0)$ whose characteristic exponents are $\pm\sqrt{c(\varepsilon)}\varepsilon^{j/2}$, with $c(\varepsilon) = -U''(0; \varepsilon)h_0''(0; \varepsilon)$, $c(0) \neq 0$ (see (8.38), (8.39)).

The stable and unstable manifolds of $(0, 0)$ coincide along a separatrix contained in the level sets $K_0(y, x; \varepsilon) = 0$. Hence, these manifolds are $\mathcal{C}^{r-2m-1}$, the same regularity as the Hamiltonian $K_0$.

When we consider also the variable $s$, the critical point $(0, 0)$ becomes a $2\pi k_0$-periodic orbit given by $\lambda_0 = \{(0, 0)\} \times \mathbb{R}/(2\pi k_0\mathbb{Z})$. The orbit $\lambda_0$ is hyperbolic with characteristic exponents $\pm\sqrt{c(\varepsilon)}\varepsilon^{j/2}$, and the splitting among the stable and unstable directions is of $O(\varepsilon^{j/2})$.

The following Proposition gives the existence of a hyperbolic periodic orbit for the full Hamiltonian (8.36) and provides quantitative estimates for the difference of this periodic orbit and its stable and unstable manifolds with those of the unperturbed system.

The main difficulty arises from the fact that the Lyapunov exponents are close to zero, as well as the splitting between stable and unstable spaces.

PROPOSITION 8.40. *With the previous notations, we have:*

(1) *The Hamiltonian system of Hamiltonian (8.36) has a periodic orbit $\lambda(\varepsilon)$ which is $\varepsilon^{m+1-j}$ close, in the $\mathcal{C}^{r-2m-4}$ sense, to the orbit $\lambda_0$. (In fact its $y$-coordinate is of order $\varepsilon^{m+1-j/2}$.)*

(2) *Given $0 < \rho < 2\pi$, we can find $\mathcal{C}^{r-2m-3}$ functions*

(8.96)                        $$\mathcal{Z}_i^{ws}, \; \mathcal{Z}_i^{wu} : [-\rho, \rho] \times \mathbb{T} \to \mathbb{R}^\pm,$$

*where $i = up, down$, so that the graph of these functions $y = \mathcal{Z}_i^{wu}(x, s)$ is a subset of a connected component of $W^{ws}_{\lambda(\varepsilon)}$, $W^{wu}_{\lambda(\varepsilon)}$, the stable and unstable manifolds in $\tilde{\Lambda}_\varepsilon$ of the periodic orbit $\lambda(\varepsilon)$ of the Hamiltonian $K(y, x, s; \varepsilon)$.*

(3) *The graphs $\mathcal{Z}_i^{\mathrm{wu}}$ are $\varepsilon^{m+1-j/2}$ close in $C^{r-2m-4}$ norm to the set $K_0(x, y; \varepsilon) = 0$. Moreover, they are given by:*

$$(8.97) \qquad y = \mathcal{Z}_{up}^{\mathrm{wu}}(x, s) = \mathcal{Y}_+(x, 0) + O_{\mathcal{C}^1}(\varepsilon^{m+1-j})$$

*where $\mathcal{Y}_\pm$ are given in (8.56). Analogous formulas hold for $\mathcal{Z}_{down}^{\mathrm{ws}}$, $\mathcal{Z}_{up}^{\mathrm{ws}}$, $\mathcal{Z}_{down}^{\mathrm{wu}}$.*

PROOF. This reduces to Theorems A and C in the paper [**FS90a**].

We consider the scaling $y = \varepsilon^{j/2}Y$, and, using Lemma 8.36, we obtain Hamiltonian (8.73).

First, we write the perturbation term in (8.73), as

$$S_1(Y, x, s; \varepsilon) = \bar{S}_1(Y, x; \varepsilon) + \tilde{S}_1(Y, x, s; \varepsilon),$$

where $\bar{S}_1$ is the averaged Hamiltonian with respect to the periodic variable $s$.

It is clear that the Hamiltonian system of Hamiltonian

$$\varepsilon^{j/2}\mathcal{K}_{\mathrm{int}} + \varepsilon^{m+1-j/2}\bar{S}_1$$

has a critical point which is $\varepsilon^{m+1-j}$ close to the origin. Then, applying the results of Theorem C and Proposition 5.1 in [**FS90a**], we obtain the existence of the periodic orbit $\lambda(\varepsilon)$, which, after the scaling $y = \varepsilon^{j/2}Y$, has the $x$ component of order $\varepsilon^{m+1-j}$, and the $y$ component of order $\varepsilon^{m+1-j/2}$.

Hamiltonian (8.73) is $2\pi k_0$-periodic in the variable $s$. Then, one can consider the time $2\pi k_0$ map $F_\varepsilon$ of this Hamiltonian, and this map has a fixed point $P_\varepsilon$, corresponding to the periodic orbit $\lambda(\varepsilon)$. We will consider also the time $2\pi k_0$ map $F_0$ of Hamiltonian $\varepsilon^{j/2}\mathcal{K}_{\mathrm{int}}$ given in (8.73).

These maps verify:

$$\|F_0 - \mathrm{Id}\|_{\mathcal{C}^{r-2m-3}} \leq O(\varepsilon^{j/2})$$
$$\|F_\varepsilon - F_0\|_{\mathcal{C}^{r-2m-4}} \leq O(\varepsilon^{m+1-j/2}),$$

and the eigenvalues of the fixed points are $1 + O(\varepsilon^{j/2})$. Then, by Theorem A in [**FS90a**], we obtain the existence of $W_{P_\varepsilon}^{\mathrm{ws,wu}}$, which are of class $\mathcal{C}^{r-2m-3}$, and $\varepsilon^{m+1-j}$ close, in the $\mathcal{C}^{r-2m-4}$ sense, to those of $K_0$. Moreover, in $[-\rho, \rho] \times \mathbb{T}$, using Lemma 8.34 and Lemma 8.39 for $E = 0$, $\delta = \varepsilon^j$ and $\nu = \varepsilon^{m+1-j}$, they can be written as a graph of the variable $Y$ over the variables $(x, s)$ verifying (8.97).

Going back to the variables $(y, x, s) = (\varepsilon^{j/2}Y, x, s)$, we obtain Proposition 8.40 for $\mathcal{Z}_i^v$.

$\square$

REMARK 8.41. Proposition 8.40 gives the stable and unstable manifolds of the periodic orbit inside $\tilde{\Lambda}_\varepsilon$ in the variables $(y, x, s)$. We can write them in the original variables $(I, \varphi, s)$ using the changes given by Proposition 8.2, Theorem 8.9 and the changes (8.32) and (8.35).

In the variables $(I, \varphi, s)$ these manifolds inside the region $\mathcal{S}^{\mathcal{R}^j}$ are given by

$$I = -k_0/l_0 + \bar{\mathcal{Z}}_i^v(\varphi, s; \varepsilon)$$

where the functions $\bar{\mathcal{Z}}_i^v(\varphi, s; \varepsilon)$ verify the same regularity properties as the functions $\mathcal{Z}_i^v$.

REMARK 8.42. Note that the results claimed in Proposition 8.40 are rather conservative and there are better results with more completed arguments. For example, applying the regularity theory of invariant manifolds in the variables $(J, \varphi, s)$, we can conclude that the invariant manifolds are as smooth as the flow of the Hamiltonian $k(J, \varphi, s; \varepsilon)$ given in (8.4), that is, $\mathcal{C}^{r-3}$. Similar improvements can be made to the dependence on parameters.

Nevertheless, for our purposes, it is more convenient to discuss the manifolds in the averaged set of coordinates since this will allow us to compare them with primary and secondary KAM tori, which are better expressed in a system of coordinates with action-angle variables. (Some references where the smooth dependence of invariant manifolds, more general than the stable and unstable, are considered rather explicitly are [**LW95, CFL03**].)

REMARK 8.43. Note that we are not even attempting to study whether $W_{\lambda(\varepsilon)}^{\text{ws}}$ and $W_{\lambda(\varepsilon)}^{\text{wu}}$ intersect transversely in $\tilde{\Lambda}_\varepsilon$ or not. Even if such questions are crucial in other approaches to diffusion, they do not play any role in our method. (See [**FS90b, FS90a**], for a discussion about the intersection of the stable and unstable manifolds in situations similar to the ones we are considering here.)

REMARK 8.44. It will be important to note that the 2-dimensional invariant manifolds of periodic orbits $W_{\lambda(\varepsilon)}^{\text{ws,wu}}$, produced by studying the dynamics on $\tilde{\Lambda}_\varepsilon$, have Lyapunov exponents that are $\varepsilon^{j/2}$ close to zero. Therefore, when we consider them in the full phase space, they are *slow* manifolds. The directions transverse to $\tilde{\Lambda}_\varepsilon$ have Lyapunov exponents of size $\tilde{\mu} = \mu + O(\varepsilon)$ (see Remark 7.4).

Hence, the manifolds $W_{\lambda(\varepsilon)}^{\text{ws,wu}}$, when considered as invariant manifolds in the whole space are only *weak* (un)stable manifolds of the periodic orbit $\lambda(\varepsilon)$. In [**LW95**] one can find a theory for these manifolds. The manifolds $W_{\lambda(\varepsilon)}^{\text{ws,wu}}$ are not the (un)stable manifolds associated in the normal hyperbolicity theory.

For our purposes, it will be useful to develop a unified notation for the 2-dimensional tori and for the stable and unstable manifolds of a periodic orbit. Note that both of them are close to a level set of the averaged energy. They will also play a similar role in the construction of transition chains later.

When $\mathcal{V} = \mathcal{T}$ is a 2-dimensional torus in $\tilde{\Lambda}_\varepsilon$, its weak (un)stable manifolds are simply $\mathcal{T}$, and therefore its associated *total* (un)stable manifolds are $W_{\mathcal{T}}^{\text{tu,ts}} = W_{\mathcal{T}}^{\text{u,s}}$.

When $\mathcal{V} = \lambda(\varepsilon)$ is a periodic orbit in $\tilde{\Lambda}_\varepsilon$, its weak (un)stable manifolds $W^{\mathrm{wu,ws}}_{\lambda(\varepsilon)}$ are the (un)stable manifolds inside the manifold $\tilde{\Lambda}_\varepsilon$. Then, we set:

$$(8.98) \qquad W^{\mathrm{tu,ts}}_{\lambda(\varepsilon)} = W^{\mathrm{u,s}}_{W^{\mathrm{wu,ws}}_{\lambda(\varepsilon)}} = \bigcup_{x \in W^{\mathrm{wu,ws}}_{\lambda(\varepsilon)}} W^{\mathrm{u,s}}_x.$$

In subsequent chapters, when we discuss transition chains, we will use the total (un)stable manifolds of different invariant objects $\tilde{\Lambda}_\varepsilon$.

REMARK 8.45. Up to now, we have applied Theorem 8.9 for a fixed constant $L$ such that the intervals $[-l/k - 2L, -l/k + 2L]$, where $-l/k \in \mathcal{R}$ are disjoint. After this, in Section 8.4 (Proposition 8.21 and Remark 8.22), Section 8.5.1 (Proposition 8.24 and Remark 8.25), sections 8.5.2, 8.5.3, 8.5.4 (Theorem 8.30, Corollary 8.31 and Remark 8.32) and Section 8.5.5 (Proposition 8.40 and Remark 8.41), we have given a complete description of the invariant objects that fill the non resonant region $\mathcal{S}^L$, and the resonant regions $\mathcal{S}^{\mathcal{R}_j}$, for $j \geq 3$ and for $j = 1, 2$ respectively.

At this moment, we have a complete description of the dynamics of the Hamiltonian flow associated to Hamiltonian $k(J, \varphi, s; \varepsilon)$ in all $\tilde{\Lambda}_\varepsilon$ except at the subsets

$$(8.99) \qquad \left( \cup_{-l/k \in \mathcal{R}} [-l/k - 2L, -l/k - L] \cup [-l/k + L, -l/k + 2L] \right) \times \mathbb{T}^2.$$

To obtain a complete description of the dynamics in (8.99), we apply Theorem 8.9 with $\tilde{L} = L/2$. The regions (8.99) are now contained in the non-resonant region corresponding to $\tilde{L}$, that is $\mathcal{S}^{\tilde{L}}$. So, the dynamics in (8.99) is also given by Proposition 8.21 and Remark 8.22, with slightly different constants.

CHAPTER 9

# The scattering map

The aim of this chapter is to define and compute the *scattering map* (also called *outer map*) $S : \tilde{\Lambda}_\varepsilon \longrightarrow \tilde{\Lambda}_\varepsilon$.

## 9.1. Some generalities about the scattering map

The following definitions and elementary facts come from [**DLS00**].

DEFINITION 9.1. *Let $\Lambda \subset M$ be a normally hyperbolic invariant manifold for a flow $\Phi_t$ in a manifold $M$.*

*Assume that $\gamma \subset W_\Lambda^s \cap W_\Lambda^u$ is a homoclinic manifold, and that the intersection of $W_\Lambda^s$ and $W_\Lambda^u$ is transverse along $\gamma$, that is*

(9.1)
$$T_z W_\Lambda^s + T_z W_\Lambda^u = T_z M, \quad \forall z \in \gamma.$$
$$T_z W_\Lambda^s \cap T_z W_\Lambda^u = T_z \gamma, \quad \forall z \in \gamma.$$

*Then, for any two points $x_\pm \in \Lambda$, we say that $x_+ = S(x_-)$, if there exists a point $z \in \gamma$ such that*

$$\mathrm{dist}(\Phi_t(z), \Phi_t(x_\pm)) \to 0, \quad for\ t \to \pm\infty.$$

Note that the definition of the scattering map $S$ depends on the homoclinic manifold $\gamma$, but we will not include this in the notation since it will not lead to confusion.

Heuristically, the scattering map associates to the asymptotic behavior in the past of homoclinic orbits in $\gamma$, the asymptotic behavior in the future. Therefore, one should think about it as an accounting device describing the homoclinic excursions.

A feature that will be crucial for us is that one can study very comfortably the intersection of stable and unstable manifolds of sets in $\Lambda$ using the scattering map independently of the topological type of the objects considered (we do not need that they are in a common system of coordinates). We will use this to discuss the existence of heteroclinic connections between primary and secondary tori or among tori of different dimensions. This is crucial for the mechanisms we introduce in this paper since we use these objects of different topologies and different dimensions to fill the gaps among KAM primary tori.

Note that the domain and the range of the scattering map associated to an intersection $\gamma$ as above are

$$\mathrm{Dom}(S) = \{x_- \in \Lambda, W^{\mathrm{u}}_{x_-} \subset \gamma\},$$
$$\mathrm{Ran}(S) = \{x_+ \in \Lambda, W^{\mathrm{s}}_{x_+} \subset \gamma\}.$$

If $\gamma$ verifies (9.1), then the dimension of $\Lambda$ and $\gamma$ are the same and we obtain that the domain and the range of the scattering map will be open (and non-empty) subsets. In the case considered in [**DLS00**], both the domain and the range of the scattering map were the whole manifold $\Lambda$. In this paper, however, we will have to pay more attention to the domains.

We recall from [**DLS00**] that the scattering map is locally well defined and as smooth as the invariant stable and unstable bundles of $\Lambda$. The reason is that if we take locally a section $\Sigma$ of $\gamma$ transversal to the flow, we can define $\Omega_\pm : \Lambda \to \Sigma$ by setting $\Omega_-(x)$ to be the (locally unique) point in $\Sigma \cap W^{\mathrm{u}}_x$ and $\Omega_+(x)$ to be the (locally unique) point in $\Sigma \cap W^{\mathrm{s}}_x$. These maps are clearly as smooth as the map $x \to W^{\mathrm{s}}_x$, $x \to W^{\mathrm{u}}_x$. The scattering map is given, locally by $S = \Omega_+ \circ \Omega_-^{-1}$.

The regularity of the maps $x \to W^{\mathrm{s,u}}_x$ is studied in great detail in [**HPS77, Fen74**]. In general, it depends on ratios of several rates of expansion. In our case, however, as we saw in Theorem 7.1, it is as smooth as the flow because the Lipschitz constant of the flow along the invariant manifold is arbitrarily close to 1 if $\varepsilon$ is small enough. Hence, for $\varepsilon$ small enough, the only constraint for the regularity of the invariant manifold is the regularity of the flow.

If we were interested in checking that the scattering map is globally defined, we would just need to check that if we continue these local definitions around a loop there is no monodromy. This follows because if $W^{\mathrm{s}}_{x_+}$ intersects non-trivially $W^{\mathrm{s}}_{\tilde{x}_+}$ and $x_+, \tilde{x}_+ \in \Lambda$, we conclude that $x_+ = \tilde{x}_+$. In this paper we may be considering situations in which $\mathrm{Dom}(S)$ is quite explicit (and happens to be contractible) and the issue of global definition of $S$ does not arise. Nevertheless in [**DLS00**] as well as in the examples in Chapter 13 the scattering map is globally defined in the unperturbed case. The explicit perturbative formulas show that there is a monodromy that tends to zero as the perturbation tends to zero.

We note however that in the applications to generate instability considered in this paper it is enough that we have possibly different scattering maps defined in open sets of $I, \varphi, s$ independent of $\varepsilon$ so that the $I$ projections are overlaping intervals. The reason is that the KAM tori are close to level sets of the $I$ variable. If the scattering map is defined in an open set as indicated above, we will show that (under appropriate non-degeneracy conditions) there are transitions chains starting in the low end of the $I$ interval and getting to the upper end of this interval. If the intervals overlap, then, we can continue the transition chains. We do not need that the scattering maps defiened in different open sets are continuations of each other.

## 9.2. The scattering map in our model: definition and computation

In this section, we will show that, in the assumptions of Theorem 4.1, the manifold $\tilde{\Lambda}_\varepsilon$ constructed in Chapter 7 has a scattering map and we will compute the leading order.

The first difficulty to define the scattering map for $\tilde{\Lambda}_\varepsilon$ in our problem is that, for $\varepsilon = 0$, its stable and unstable manifolds coincide: $W^s_{\tilde{\Lambda}} = W^u_{\tilde{\Lambda}}$. We will show that under hypothesis **H4** of Theorem 4.1 the stable and unstable invariant manifolds of $\tilde{\Lambda}_\varepsilon$ have a transversal homoclinic intersection for $0 < |\varepsilon| < \varepsilon^*$. (This will just amount to verifying that in the situation considered here, the conditions in first order perturbation theory **H4** are enough to guarantee the existence of a transversal intersection associated to the critical point $\tau^*$ of the map (4.5).)

The transversal intersection associated to the critical point $\tau^*$ will happen for the points in $H_- \subset \tilde{\Lambda}_\varepsilon$, where $H_-$ is defined in (4.4). Hence, it will be possible to define a scattering map $S$ with $H_- \subset \mathrm{Dom}(S)$. We will denote by $H_+ = S(H_-)$, so we have that $H_+ \subset \mathrm{Ran}(S)$.

Both the verification of the existence of the scattering map and explicit computations will be accomplished by first order perturbation theory of the stable and unstable manifolds adapting the calculations carried out in Section 7.1. Again we emphasize that the scattering map depends on the choice of a homoclinic intersection. In general, there will more than one and each one of them has a scattering map.

It is convenient to take advantage of the exact symplectic structure of the problem. We will see that the existence of homoclinic intersections will be given, by first order perturbation theory, by the zeros of a directional derivative of a single function called the *Poincaré function* or the *Melnikov potential*, which was written explicitly in (4.3), and introduced in [**DG00**].

As we will see, the Poincaré function not only appears in the existence of intersections for $W^s_{\tilde{\Lambda}_\varepsilon}$, $W^u_{\tilde{\Lambda}_\varepsilon}$ but also in the calculation of the scattering map (see (9.9)). This, in turn, will help us with the calculations of the existence of transition chains in Section 10.3. The hypothesis **H4** and, hence, the other subsequent hypothesis **H5"**, **H5"'** are expressed in terms of the Poincaré function.

We recall from (4.3) that the definition of the Poincaré function is:

$$\text{(9.2)} \quad \mathcal{L}(I,\varphi,s) := -\int_{-\infty}^{+\infty} \Big( h(p_0(\sigma), q_0(\sigma), I, \varphi + I\sigma, s + \sigma) \\ -h(0,0,I,\varphi + I\sigma, s + \sigma)\Big) d\sigma,$$

where $(p_0(t), q_0(t))$ is the parameterization (6.1) of the separatrix to the saddle point $(0,0)$ of the pendulum $P_\pm(p,q) = \pm(p^2/2 + V(q))$ with characteristic exponent $\mu = \sqrt{-V''(0)} > 0$.

Note that the integrand in (9.2) converges exponentially fast as $\sigma$ tends to $\pm\infty$ and that the convergence is uniform for bounded intervals in $I, \varphi, s$. Hence, the integral can be differentiated under the integral.

The following Proposition 9.2 establishes the existence of transverse intersections for $W^{\mathrm{s}}(\tilde{\Lambda}_\varepsilon)$ and $W^{\mathrm{u}}(\tilde{\Lambda}_\varepsilon)$ along a manifold $\tilde{\gamma}_\varepsilon$ under the assumption **H4** of Theorem 4.1 and provides with perturbative formulas for the scattering map associated to these intersections.

PROPOSITION 9.2. *Given* $(I, \varphi, s) \in [I_-, I_+] \times \mathbb{T} \times \mathbb{T}$, *assume that the function* $\tau \in \mathbb{R} \mapsto \mathcal{L}(I, \varphi - I\tau, s - \tau)$ *has a non-degenerate critical point at* $\tau = \tau^*(I, \varphi, s)$. *(By the implicit function theorem, the function* $\tau^*$ *is smooth).*

*Then, for* $\varepsilon$ *small enough, there exists a locally unique point* $\tilde{z}$ *of the form*

(9.3)
$$\tilde{z}(I, \varphi, s; \varepsilon) = \tilde{z}(\tau^*(I, \varphi, s), I, \varphi, s; \varepsilon) = (p_0(\tau) + \mathrm{O}(\varepsilon), q_0(\tau) + \mathrm{O}(\varepsilon), I, \varphi, s),$$

*such that* $W^{\mathrm{s}}(\tilde{\Lambda}_\varepsilon) \pitchfork W^{\mathrm{u}}(\tilde{\Lambda}_\varepsilon)$ *at* $\tilde{z}$, *that is,*

$$\tilde{z} \in W^{\mathrm{s}}(\tilde{\Lambda}_\varepsilon) \cap W^{\mathrm{u}}(\tilde{\Lambda}_\varepsilon) \text{ and } T_{\tilde{z}}W^{\mathrm{s}}(\tilde{\Lambda}_\varepsilon) + T_{\tilde{z}}W^{\mathrm{u}}(\tilde{\Lambda}_\varepsilon) = T_{\tilde{z}}\mathcal{M},$$

*where* $\mathcal{M} = \mathbb{R} \times \mathbb{T} \times [I_-, I_+] \times \mathbb{T} \times \mathbb{T}$.

*In particular, there exist unique points*

$$\tilde{x}_\pm = \tilde{x}_\pm(I, \varphi, s; \varepsilon) = (0, 0, I, \varphi, s) + \mathrm{O}_{\mathcal{C}^1}(\varepsilon) \in \tilde{\Lambda}_\varepsilon$$

*such that*

(9.4)                $$|\Phi_{t,\varepsilon}(\tilde{z}) - \Phi_{t,\varepsilon}(\tilde{x}_\pm)| \leq \mathrm{cte.}\, e^{-\mu|t|/2} \text{ for } t \to \pm\infty.$$

*Moreover, expressing the points* $\tilde{x}_\pm = \tilde{\mathcal{F}}(I_\pm, \varphi_\pm, s_\pm; \varepsilon)$ *in terms of the parameterization (7.1) of* $\tilde{\Lambda}_\varepsilon$ *given in Theorem 7.1, the following formulas hold:*

$$I(\tilde{x}_\pm) = I + \mathrm{O}_{\mathcal{C}^1}(\varepsilon), \qquad \varphi(\tilde{x}_\pm) = \varphi + \mathrm{O}_{\mathcal{C}^1}(\varepsilon), \qquad s(\tilde{x}_\pm) = s,$$

*and*

$$I(\tilde{x}_+) - I(\tilde{x}_-) = \varepsilon \frac{\partial \mathcal{L}}{\partial \varphi}(I, \varphi - I\tau, s - \tau) + \mathrm{O}_{\mathcal{C}^1}(\varepsilon^{1+\varrho}),$$

*where* $\tau$ *is given again by* $\tau = \tau^*(I, \varphi, s)$, *and* $\varrho > 0$.

REMARK 9.3. We recall that the first part of hypothesis **H4** in Theorem 4.1 is precisely that the hypothesis of Proposition 9.2 occurs with uniform constants for a non-empty open set $H_- \subset \tilde{\Lambda}_\varepsilon$.

Hence, for positive $\varepsilon$ we have transveal intersections and we can define an scattering map with domain $H_-$. Given the explicit computations of the scattering map later, the second part of the hypothesis **H4** imply that in this domain, the scattering map increases the $I$ (resp. decreases).

To complete the proof of the existence of transition chains, we will introduce later some explicit non-degeneracy conditions of the scattering map (roughly, that the images of KAM tori intersect other KAM tori transversally).

PROOF. The basic idea of the proof of Proposition 9.2 is very standard in Melnikov theory. We start by observing that in the unperturbed case, the stable and unstable manifolds are characterized as the set $\{P_\pm = 0\}$ where $P_\pm$ is the pendulum part of the unperturbed Hamiltonian. Hence, outside of the critical points of $P_\pm$, we can construct a system of coordinates given by a point in the homoclinic intersection and a value for $P_\pm$. On the other hand, we observe that since $P_\pm$ is a conserved quantity in the unperturbed system, $\frac{d}{dt}P_\pm$ is small and can be computed perturbatively. Hence, the equation of an invariant manifold can be solved perturbatively and the fist term is an integral over the perturbed quantities. Under assumptions of non-degeneracy, this first order calculation is enough to conclude the desired results of existence of transverse intersections (using the implicit function theorem). Similar procedures have been implemented in [**DG00**]. See also [**Tre94**]. A different method to prove persistence of intersections—not necessarily transversal—is in [**Eli94**].

We proceed to give details of the calculation.

Consider the point

$$\tilde{z}_0 = \tilde{z}_0(\tau, I, \varphi, s) := (p_0(\tau), q_0(\tau), I, \varphi, s)$$

of the unperturbed homoclinic 4-dimensional manifold $\tilde{\gamma}$ for the unperturbed system given in (6.6). The line

$$N = \left\{ \tilde{z}_0 + u \left( \frac{\partial P_\pm}{\partial p}(p_0(\tau), q_0(\tau)), \frac{\partial P_\pm}{\partial q}(p_0(\tau), q_0(\tau)), 0, 0, 0 \right), u \in \mathbb{R} \right\}$$

where $P_\pm$ is the pendulum part of the unperturbed Hamiltonian, is normal to $\tilde{\gamma}$ at $\tilde{z}_0$: $N \pitchfork_{\tilde{z}_0} \tilde{\gamma}$. Since $W_{\tilde{\Lambda}_\varepsilon}^{\mathrm{s,u,loc}}$ is $C^1$ $\varepsilon$-close to $\tilde{\gamma}$, it follows that $N$ intersects transversally $W_{\tilde{\Lambda}_\varepsilon}^{\mathrm{s,u,loc}}$ at a locally unique $\tilde{z}^{\mathrm{s,u}} = \tilde{z}^{\mathrm{s,u}}(\tau, I, \varphi, s; \varepsilon) \in N$, $\varepsilon$-close to $\tilde{z}_0$. Notice that $\tilde{z}^{\mathrm{s,u}}$ has the form

$$\tilde{z}^{\mathrm{s,u}} = (p_0(\tau) + \mathrm{O}_{\mathcal{C}^1}(\varepsilon), q_0(\tau) + \mathrm{O}_{\mathcal{C}^1}(\varepsilon), I, \varphi, s).$$

Therefore, by the implicit function theorem, $W^{\mathrm{s}}(\tilde{\Lambda}_\varepsilon)$, $W^{\mathrm{u}}(\tilde{\Lambda}_\varepsilon)$ will have a non-empty intersection close to $\tilde{z}_0$ when

(9.5)                     $$DP_\pm(p_0(\tau), q_0(\tau))(\tilde{z}^{\mathrm{u}} - \tilde{z}^{\mathrm{s}}) = 0,$$

where $DP_\pm(p, q)(\tilde{z}^{\mathrm{u}} - \tilde{z}^{\mathrm{s}})$ is a shorthand for

$$(DP_\pm(p, q), 0, 0, 0)(\tilde{z}^{\mathrm{u}} - \tilde{z}^{\mathrm{s}}).$$

To provide a more easily computable formula for (9.5), we first notice that, for $\varepsilon$ small enough, as a consequence of the normal hyperbolicity of $\tilde{\Lambda}_\varepsilon$, there exist unique points $\tilde{x}_\pm = \tilde{x}_\pm(I, \varphi, s; \varepsilon) \in \tilde{\Lambda}_\varepsilon$ such that $\Phi_{t,\varepsilon}(\tilde{z}^{\mathrm{s,u}}) - \Phi_{t,\varepsilon}(\tilde{x}_\pm) \to 0$ exponentially with an exponent bounded away from zero as in (9.4) for $t \to \pm\infty$, and that $\Phi_{t,\varepsilon}(\tilde{z}^{\mathrm{s,u}}) - \Phi_{t,0}(\tilde{z}_0) = \mathcal{O}_{\mathcal{C}^1}(\varepsilon)$, uniformly in $\varepsilon$ for $t \to \pm\infty$. In particular, $\tilde{x}_\pm = (0, 0, I, \varphi, s) + \mathcal{O}_{\mathcal{C}^1}(\varepsilon)$ and

$$DP_\pm \left( p_0(\tau + t), q_0(\tau + t) \right) \left( \Phi_{t,\varepsilon}(\tilde{z}^{\mathrm{s,u}}) \right) \longrightarrow 0 \text{ for } t \to \pm\infty.$$

We apply the fundamental theorem of calculus to

$$t \mapsto DP_\pm(p_0(\tau + t), q_0(\tau + t)) \left( \Phi_{t,\varepsilon}(\tilde{z}^{\mathrm{s,u}}) \right),$$

to get

$$
\begin{aligned}
DP_\pm(p_0(\tau), q_0(\tau))(\tilde{z}^{\mathrm{s,u}}) &= \varepsilon \int_{\pm\infty}^0 \{P_\pm, h\} \left( \Phi_{\sigma,\varepsilon}(\tilde{z}^{\mathrm{s,u}}) \right) d\sigma \\
&= \varepsilon \int_{\pm\infty}^0 \{P_\pm, h\} \left( \Phi_{\sigma,0}(\tilde{z}_0) \right) d\sigma + \mathcal{O}_{\mathcal{C}^1}(\varepsilon^2),
\end{aligned}
$$

where $\{P_\pm, h\} = \partial_q P_\pm \, \partial_p h - \partial_p P_\pm \, \partial_q h$ is the Poisson bracket of the functions $P_\pm$ and $h$. Subtracting the expressions above and taking into account the uniform convergence of the integrands, we get

$$DP_\pm(p_0(\tau), q_0(\tau))(\tilde{z}^{\mathrm{u}} - \tilde{z}^{\mathrm{s}}) = \varepsilon \frac{\partial L}{\partial \tau}(\tau, I, \varphi, s) + \mathcal{O}_{\mathcal{C}^1}(\varepsilon^2)$$

where $L(\tau, I, \varphi, s)$ is given by

$$
\begin{aligned}
\text{(9.6)} \qquad L(\tau, I, \varphi, s) = - \int_{-\infty}^{+\infty} &\Big( h(p_0(\tau + \sigma), q_0(\tau + \sigma), I, \varphi + I\sigma, s + \sigma) \\
&- h(0, 0, I, \varphi + I\sigma, s + \sigma) \Big) d\sigma.
\end{aligned}
$$

From the expression of $L(\tau, I, \varphi, s)$, it is immediate to check that

$$L(\tau + t, I, \varphi + It, s + t) = L(\tau, I, \varphi, s) \text{ for any } t \in \mathbb{R}.$$

Therefore, writing $t = -\tau$, we arrive at

$$L(\tau, I, \varphi, s) = L(0, I, \varphi - I\tau, s - \tau) = \mathcal{L}(I, \varphi - I\tau, s - \tau),$$

where $\mathcal{L}$ is the Melnikov potential defined in (4.3). Then, equation (9.5) is equivalent to

$$\varepsilon \frac{\partial \mathcal{L}}{\partial \tau}(I, \varphi - I\tau, s - \tau) + \mathcal{O}_{\mathcal{C}^1}(\varepsilon^2) = 0.$$

By the implicit function Theorem, non-degenerate critical points $\tau$ of the function $\tau \mapsto \mathcal{L}(I, \varphi - I\tau, s - \tau)$ give rise, for $\varepsilon$ small enough, to transverse intersections of the stable and unstable manifolds of $\tilde{\Lambda}_\varepsilon$ along points $\tilde{z} = \tilde{z}(\tau, I, \varphi, s; \varepsilon)$ of the form (9.3).

By the implicit function theorem, we can find a function $\tau^*$ defined in an open set where $\tau^*(I; \varphi, s)$ is a critical point of $\tau \mapsto \mathcal{L}(I, \varphi - I\tau, s - \tau)$.

To finish the proof of Proposition 9.2, we consider the expression of the points $\tilde{x}_\pm = \tilde{\mathcal{F}}(I_\pm, \varphi_\pm, s_\pm; \varepsilon)$ in terms of the parameterization (7.1) of $\tilde{\Lambda}_\varepsilon$ given in Theorem 7.1. Since we already know the existence of $\tilde{z}$ given in (9.3) such that (9.4) holds, it is clear that

$$I(\tilde{x}_\pm) = I + \mathcal{O}_{\mathcal{C}^1}(\varepsilon), \qquad \varphi(\tilde{x}_\pm) = \varphi + \mathcal{O}_{\mathcal{C}^1}(\varepsilon), \qquad s(\tilde{x}_\pm) = s,$$

and to finish the proof of Proposition 9.2 it only remains to obtain the formula for $I(\tilde{x}_+) - I(\tilde{x}_-)$.

We apply now the Fundamental Theorem of Calculus to

$$t \mapsto I\left(\Phi_{t,\varepsilon}(\tilde{x}_\pm)\right) - I\left(\Phi_{t,\varepsilon}(\tilde{z})\right),$$

to get

$$I(\tilde{x}_\pm) - I(\tilde{z}) = \varepsilon \int_{\pm\infty}^{0} \left(\{I,h\}\left(\Phi_{\sigma,\varepsilon}(\tilde{x}_\pm)\right) - \{I,h\}\left(\Phi_{\sigma,\varepsilon}(\tilde{z})\right)\right) d\sigma,$$

where $\{I,h\} = \partial_\varphi I\, \partial_I h - \partial_I I\, \partial_\varphi h = -\partial_\varphi h$ is the Poisson bracket of the functions $I$ and $h$. Subtracting the expressions above, we get

$$I(\tilde{x}_+) - I(\tilde{x}_-) = \varepsilon \int_{0}^{+\infty} \frac{\partial h}{\partial \varphi}(\Phi_{\sigma,\varepsilon}(\tilde{z})) - \frac{\partial h}{\partial \varphi}(\Phi_{\sigma,\varepsilon}(\tilde{x}_+)) d\sigma$$

$$+\varepsilon \int_{-\infty}^{0} \frac{\partial h}{\partial \varphi}(\Phi_{\sigma,\varepsilon}(\tilde{z})) - \frac{\partial h}{\partial \varphi}(\Phi_{\sigma,\varepsilon}(\tilde{x}_-)) d\sigma.$$

We already know that

$$\Phi_{t,\varepsilon}(\tilde{z}) = \Phi_{t,0}(\tilde{z}_0) + \mathcal{O}_{\mathcal{C}^1}(\varepsilon), \quad \forall t \in \mathbb{R}.$$

Taking $c_2$ sufficiently small (but independent of $\varepsilon$) and using Gronwall inequality we have, for $-c_2\,|\log \varepsilon| \le t \le c_2\,|\log \varepsilon|$

$$\Phi_{t,\varepsilon}(\tilde{x}_\pm) = \Phi_{t,0}(\tilde{x}_0) + \mathcal{O}_{\mathcal{C}^1}(\varepsilon^{\varrho_1}),$$

for some $\varrho_1 > 0$.

From equation (9.4), we deduce that there exists a constant $c_1 > 0$,

$$\left| \int_{\pm c_2 |\log \varepsilon|}^{\pm\infty} \left( \frac{\partial h}{\partial \varphi}(\Phi_{\sigma,\varepsilon}(z)) - \frac{\partial h}{\partial \varphi}(\Phi_{\sigma,\varepsilon}(x_\pm)) \right) d\sigma \right|$$

$$\le c_1 e^{-\mu c_2 |\log \varepsilon|/2} = \mathcal{O}_{\mathcal{C}^1}(\varepsilon^{\varrho_2}),$$

and, since for the unperturbed system we have the same kind of behavior, we can conclude finally that, for some $\varrho > 0$

$$I(\tilde{x}_+) - I(\tilde{x}_-)$$

$$= -\varepsilon \int_{-c_2|\log \varepsilon|}^{+c_2|\log \varepsilon|} \left( \frac{\partial h}{\partial \varphi}(q_0(\tau + \sigma), p_0(\tau + \sigma), I, \varphi + I\sigma, s + \sigma) \right.$$
$$\left. - \frac{\partial h}{\partial \varphi}(0, 0, I, \varphi + I\sigma, s + \sigma) \right) d\sigma + O_{\mathcal{C}^1}(\varepsilon^{1+\varrho})$$

$$= -\varepsilon \int_{-\infty}^{+\infty} \left( \frac{\partial h}{\partial \varphi}(q_0(\tau + \sigma), p_0(\tau + \sigma), I, \varphi + I\sigma, s + \sigma) \right.$$
$$\left. - \frac{\partial h}{\partial \varphi}(0, 0, I, \varphi + I\sigma, s + \sigma) \right) d\sigma + O_{\mathcal{C}^1}(\varepsilon^{1+\varrho})$$

$$= -\varepsilon \int_{-\infty}^{+\infty} \left( \frac{\partial h}{\partial \varphi}(q_0(r), p_0(r), I, \varphi + Ir - I\tau, s + r - \tau) \right.$$
$$\left. - \frac{\partial h}{\partial \varphi}(0, 0, I, \varphi + Ir - I\tau, s + r - \tau) \right) dr + O_{\mathcal{C}^1}(\varepsilon^{1+\varrho})$$

$$= -\varepsilon \frac{\partial \mathcal{L}}{\partial \varphi}(I, \varphi - I\tau, s - \tau) + O_{\mathcal{C}^1}(\varepsilon^{1+\varrho}).$$

$\square$

REMARK 9.4. Observe that, by the definition of $\tau^*(I, \varphi, s)$, the function $\mathcal{L}(I, \varphi - I\tau^*(I, \varphi, s), s - \tau^*(I, \varphi, s))$ is a solution of the equation

$$I\partial_\varphi f(I, \varphi, s) + \partial_s f(I, \varphi, s) = 0,$$

or, equivalently, there exists a function $\mathcal{L}^*(I, \theta)$, that we will call the reduced Poincaré function, defined by

$$(9.7) \qquad \mathcal{L}(I, \varphi - I\tau^*(I, \varphi, s), s - \tau^*(I, \varphi, s)) := \mathcal{L}^*(I, \varphi - Is).$$

If we assume hypothesis **H4**, again by the definition of $\tau^*$, for $(I, \varphi, s) \in H_-$ (see (4.4)), the function $\mathcal{L}^*$ verifies

$$\frac{\partial \mathcal{L}}{\partial x}(I, \varphi - I\tau^*(I, \varphi, s), s - \tau^*(I, \varphi, s)) = \frac{\partial \mathcal{L}^*}{\partial x}(I, \varphi - Is),$$

for $x = I, \varphi, s$.

We denote by $H_-^*$ the domain of $\mathcal{L}^*$. Clearly

$$(9.8) \quad H_-^* = \{(I, \theta) : \ \theta = I - \varphi s, \ (I, \varphi, s) \in H_-\} = \cup_{I \in (I_-, I_+)} \{I\} \times \mathcal{J}_I^*.$$

Finally, from (9.7), the scattering map $S$ on $\tilde{\Lambda}_\varepsilon$ can be computed

$$S : H_- \subset \tilde{\Lambda}_\varepsilon \longrightarrow \tilde{\Lambda}_\varepsilon$$

(9.9)
$$(I, \varphi, s) \longmapsto (I - \varepsilon \frac{\partial \mathcal{L}^*}{\partial \varphi}(I, \varphi - Is) + \mathrm{O}_{\mathcal{C}^1}(\varepsilon^{1+\varrho}), \varphi + \mathrm{O}_{\mathcal{C}^1}(\varepsilon), s).$$

Note that the scattering map is $\mathrm{O}(\varepsilon)$ close to the identity. We have computed the leading term of the $I$ component but not of the $\varphi$ component. This will be enough for our purposes since we will be mainly concerned with the action of the scattering map on KAM tori which are close to level sets of $I$. Hence, changes in $\varphi$ have a much smaller effect. The calculation of the $I$ variable is greatly facilitated by the fact that $I$ is a slow variable and its change can be obtained just by the fundamental theorem of calculus. In the paper [**DLS04**] there are some more detailed calculations of this scattering map.

CHAPTER 10

# Existence of transition chains

The goal of this chapter is to construct transition chains—see Definition 10.2 in Section 10.1—along the manifold $\tilde{\Lambda}_\varepsilon$. In particular, these transition chains will traverse the resonant regions. Later, in Chapter 11 we will show that, once we have a transition chain, there are orbits that follow it closely. Hence, there are orbits that traverse the resonant regions as claimed in Theorem 4.1.

Transition chains were an important ingredient of the method in [**Arn64**]. The main novelty here is that we introduce two new ingredients in the transition chains. We allow the transition chains to include secondary tori and periodic orbits, as well as the more customary primary tori.

In Section 10.2, we give explicit conditions on the perturbations considered in (3.3) that imply that there are transition chains that overcome the large gap problem. These conditions are part of **H4** and **H5** in Theorem 4.1.

First, in Section 10.2 we show how the scattering map, constructed in Chapter 9, can be used as a tool to discuss heteroclinic intersections of objects of different topological types. Roughly—see Lemma 10.4—two submanifolds $\mathcal{V}_1$, $\mathcal{V}_2$ of $\tilde{\Lambda}_\varepsilon$ have a transverse heteroclinic connection, if $S(\mathcal{V}_1)$ is transversal to $\mathcal{V}_2$ as submanifolds of $\tilde{\Lambda}_\varepsilon$. This provides us with an alternative to the customary Melnikov calculations, which require common systems of coordinates for both objects.

Recall that in Chapter 9 we have obtained an explicit expression (9.9) for the scattering map. In Propositions 8.21, 8.24, and Corollary 8.31 we have obtained explicit expressions (depending on the proximity to resonances) for the KAM tori, both primary or secondary, in $\tilde{\Lambda}_\varepsilon$. In Proposition 8.40, we have studied the weak invariant manifolds in $\tilde{\Lambda}_\varepsilon$ of periodic orbits associated to double resonances. (See Figure 8.1 for a representation of the results obtained so far). Roughly speaking, all these objects are given very approximately by the level sets of the averaged Hamiltonian.

In Lemma 10.7 we will prove a general result that allows to verify transverse intersection in $\tilde{\Lambda}_\varepsilon$ of $S(\mathcal{V}_1)$ and $\mathcal{V}_2$, when $\mathcal{V}_1$, $\mathcal{V}_2$, submanifolds of $\tilde{\Lambda}_\varepsilon$, are given as level sets of some function. In Lemmas 10.8, 10.11, 10.14 we will apply Lemma 10.7 to the cases of primary or secondary tori or (un)stable manifolds of periodic orbits.

Finally, in Section 10.3 we show that all the gaps between the objects mentioned are close enough in order to obtain transition chains between them.

A pictorial representation of the results of this chapter is in Figure 10.1. For simplicity of the pictorial representation we have depicted only the case where the stronger version of **H4** discussed in Remark 4.3 holds.

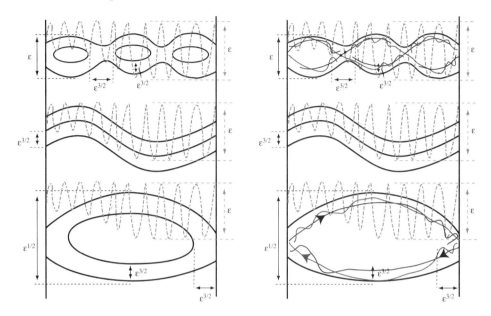

FIGURE 10.1. Schematic illustration of the main invariant objects in $\tilde{\Lambda}_\varepsilon$ studied in Chapter 8 as well as their images (dashed) under the scattering map.

Figure 10.1 depicts a 2-dimensional surface of section of the 3-dimensional manifold $\tilde{\Lambda}_\varepsilon$. We have represented in continuous lines sections of some primary and secondary tori on the left and primary tori and (un)stable manifolds of periodic orbits on the right.

The dashed lines represent the image of some primary tori under the scattering map. They intersect transversally in $\tilde{\Lambda}_\varepsilon$ other primary tori, secondary tori or (un)stable manifolds of periodic orbits.

The transition chains will be constructed by taking chains of objects $\mathcal{V}_i$ such that

$$S(\mathcal{V}_i) \pitchfork_{\tilde{\Lambda}_\varepsilon} \mathcal{V}_{i+1}.$$

Since the scattering map (9.9) moves amounts $\mathrm{O}(\varepsilon)$ in a direction somewhat different to the level sets—which have gaps $\mathrm{O}(\varepsilon^{3/2})$—it is clear that one can construct transition chains from the bottom of the figure to the top.

REMARK 10.1. We think that the incorporation of secondary tori and tori of lower dimension in the transition chains is quite natural. The resonant regions are devoid of primary KAM tori. Hence, if there are geometric mechanisms that produce diffusion across resonances, it is clear that they have to involve other geometric objects besides the customary primary KAM tori. Among all the objects that appear, the secondary tori and the periodic orbits are the most prominent, as witnessed by the fact that, very early in the numerical explorations, they were identified and given rather colorful names (e.g. islands, chaotic seas, X-points etc.) and discussed in great detail.

## 10.1. Transition chains

We will find it convenient to define transition chains as:

DEFINITION 10.2. *Let $\mathcal{H}$ be a vector field, and $N \in \mathbb{N} \cup \{\infty\}$.*

*A transition chain consists of a sequence $\{\mathcal{T}_i\}_{i=0}^N$ of $C^1$ partially hyperbolic tori, invariant by $\mathcal{H}$ such that:*

i) *The motion on each of the tori $\mathcal{T}_i$ is topologically conjugated to a transitive rotation on a torus.*

ii) *There exist orbits $\gamma_i \subset W_{\mathcal{T}_{i-1}}^{\mathrm{tu}} \cap W_{\mathcal{T}_i}^{\mathrm{ts}}$, where $W_{\mathcal{T}}^{\mathrm{tu,ts}}$ are defined in (8.98).*

iii) *The intersection $W_{\mathcal{T}_{i-1}}^{\mathrm{tu}} \cap W_{\mathcal{T}_i}^{\mathrm{ts}}$ is transversal along the orbit $\gamma_i$.*

We emphasize that we are not assuming that the tori have the same dimension, or that they are homotopic.

As the tori are assumed to be partially hyperbolic, there exist $\delta_i$ such that

$$
\begin{aligned}
\mathrm{dist}(\gamma_i(t), \mathcal{T}_{i-1}) &\leq C e^{-\delta_i |t|} \quad t \leq 0, \\
\mathrm{dist}(\gamma_i(t), \mathcal{T}_i) &\leq C e^{-\delta_i |t|} \quad t \geq 0.
\end{aligned}
$$
(10.1)

We are not assuming that the $\delta_i$ are bounded away from zero uniformly in $i$. In the case that $\mathcal{H}$ depends on some parameter $\varepsilon$, we are not assuming the $\delta_i$ are bounded away from zero uniformly in $\varepsilon$.

REMARK 10.3. It is conceivable that the diffusion based on objects of different dimension has different quantitative properties.

One can argue heuristically that the lower the dimension of the orbit, the faster the transition time since the "ergodization time" is smaller the smaller the dimension of the tori. In this paper, however, we have not investigated these quantitative properties.

## 10.2. The scattering map and the transversality of heteroclinic intersections

.

We first turn to proving the result about characterizing the existence of transverse heteroclinic intersections among two different foliations of invariant manifolds of $\tilde{\Lambda}_\varepsilon$ in terms of their transversality under the scattering map provided that some non-degeneracy condition holds. This will be a general result that holds for normally hyperbolic invariant manifolds with transversal homoclinic intersections. We will prove it in this generality.

For our applications, of course, the normally hyperbolic invariant manifold will be the $\tilde{\Lambda}_\varepsilon$ produced in Chapter 7 and the foliation will be the foliation by the level sets of the averaged energy. The level sets of the averaged energy are very close to the invariant tori, so we will get transversality between the foliation of invariant tori and its image under the scattering map.

The non-degeneracy condition (Hypothesis **H5** of Theorem 4.1) will be just that the first order perturbative calculation of the angle of intersection between foliations does not vanish. We will provide explicit formulas for this first order calculation. Since we have already developed the first order calculation for the scattering map and for the foliations, the calculation of the intersections are sort of straightforward. Of course, since the calculations of the averaged energy are different in the non-resonant region and in the different regions of the resonant region (we have to distinguish between first order resonances, second order resonances and different regions of proximity), we will get different analytic expressions for the intersection of the foliations in each of these cases. The different parts of Hypothesis **H5** correspond to explicit calculations of the angle of the intersection in the non-resonant region, near the primary tori and near the secondary tori.

The fact that one should expect this transversality generically in our applications is intuitively clear because the foliation by the averaged energy depends only on the system in a neighborhood of $\tilde{\Lambda}_\varepsilon$ whereas the scattering map is affected by perturbations along the homoclinic intersection.

It is also important to recall that we will be considering only the regions when the scattering map moves by an amount bounded from below by $A\varepsilon$ for some constant $A > 0$. That is, we will be considering only heteroclinic intersections that move by a somewhat large amount. This is in contrast with many of the applications of Melnikov method in the literature. Usually, the applications of the Melnikov method lead to constructing the heteroclinic intersections as a consequence of the existence of transversal homoclinic ones. The scattering map can compute the heteroclinic intersections directly. This is why our transversality conditions are not non-degeneracy of the zeros of the Melnikov function $\frac{\partial \mathcal{L}}{\partial \varphi}(I, \varphi - I\tau^*, s - \tau^*)$ in (4.6)—as is customary in the papers based on the Melnikov method. We just need that the scattering map moves the energy foliation so that it is transversal to itself. The same observation happened in [**DLS00, DLS05**]. In those papers, it was only necessary to assume that the leading term of the Melnikov function was non-trivial. It was not assumed that it had a non-degenerate zero.

LEMMA 10.4. *Let $\tilde{\Lambda}$ be a normally hyperbolic invariant manifold. Assume that $W^s_{\tilde{\Lambda}} \pitchfork W^u_{\tilde{\Lambda}}$ along a manifold $\tilde{\gamma}$, and denote by $S$ the scattering map associated to this intersection $\tilde{\gamma}$.*

*Let $\mathcal{V}_1, \mathcal{V}_2 \subset \tilde{\Lambda}$ be $C^1$ submanifolds of $\tilde{\Lambda}$, and assume that $S(\mathcal{V}_1) \pitchfork_{\tilde{\Lambda}} \mathcal{V}_2$. (In particular $\mathcal{V}_1 \cap H_- \neq \emptyset$ and $\mathcal{V}_2 \cap H_+ \neq \emptyset$.)*

*Then, there exists a heteroclinic trajectory $\gamma_1$ such that $\gamma_1 \subset W^u_{\mathcal{V}_1} \pitchfork W^s_{\mathcal{V}_2}$.*

PROOF. By the definition of the scattering map, we have $W_{\mathcal{V}_1}^{\mathrm{u}} \cap \tilde{\gamma} = W_{S(\mathcal{V}_1)}^{\mathrm{s}} \cap \tilde{\gamma}$.

By the assumption of transversality of $S(\mathcal{V}_1)$ and $\mathcal{V}_2$ in $\tilde{\Lambda}$, we obtain $W_{S(\mathcal{V}_1)}^{\mathrm{s}} \pitchfork_{W^{\mathrm{s}}(\tilde{\Lambda})} W_{\mathcal{V}_2}^{\mathrm{s}}$, and therefore, $W_{S(\mathcal{V}_1)}^{\mathrm{s}} \pitchfork_{\tilde{\gamma}} W_{\mathcal{V}_2}^{\mathrm{s}}$. Hence,

$$W_{\mathcal{V}_1}^{\mathrm{u}} \pitchfork_{\tilde{\gamma}} W_{\mathcal{V}_2}^{\mathrm{s}}.$$

Since $W_{\tilde{\Lambda}}^{\mathrm{s}} \pitchfork W_{\tilde{\Lambda}}^{\mathrm{u}}$ along $\tilde{\gamma}$, we obtain the desired result.          □

We now formulate and prove Lemma 10.7, that will allow us to verify the conditions of Lemma 10.4 in the case that the manifolds are close to level sets of a function. This Lemma will be useful for us since the objects we have considered before (the primary and secondary tori, the weak stable and unstable manifolds of periodic orbits) are close to being level sets of the averaged Hamiltonian as we have established in Chapter 8. (See specially Proposition 8.21, Proposition 8.24, Theorem 8.30 and Proposition 8.40.)

The precise application of Lemma 10.4 to the case that the manifolds are flat—primary KAM tori far from resonance—is done in Lemma 10.8, and for non-flat manifolds—primary and secondary tori close to the resonance or weak invariant manifolds of the lower dimensional tori—is done in Lemma 10.11 for the case of a resonance of order 1 and in Lemma 10.14 in the case of a resonance of order 2.

Lemma 10.7 is designed to deal in a unified way with the different type of tori that appear in our problem. That is, the rather flat tori that appear in the non-resonant region or in the resonant regions of order 3 or bigger, and the curved tori that appear near resonances of order 1 or 2.

As we have argued in Remark 8.29 these two types of tori have different quantitative properties and this leads to the fact that the leading terms in the asymptotics of the angle of intersection between a torus and the scattering map of another one will be different depending on which class of tori we are considering. The different parts of Hypothesis **H5** will be that the expressions for the angle of intersection in the different types of tori is non-zero.

REMARK 10.5. The case of intersections of flat tori is significantly easier and can be dealt with other methods. Indeed, the study of intersections of flat tori is significantly easier than the study of [**DLS00**, Lemma 4.21]. In the case of [**DLS00**] the tori were flat but also presented a phase shift which does not appear in our case. However, in this paper, we will present the general approach that works in all cases.

Lemma 10.7 considers a foliation $\mathcal{F}_F$ whose leaves are the level sets of a function $F$:

$$L_E^F = \{(I, \varphi, s) \in (I_-, I_+) \times \mathbb{T}^2, \ F(I, \varphi, s; \varepsilon) = E\}, \ E \in (E_1, E_2)$$

and that are also parameterized as

$$L_E^F = \{(I, \varphi, s) \in (I_-, I_+) \times \mathbb{T}^2, \ I = \lambda_E(\varphi, s; \varepsilon)\},$$

and it gives criteria to establish that their leaves intersect transversally their images under the scattering map $S$. Note that these images are contained in the leaves of the function $F \circ S^{-1}$.

REMARK 10.6. Given two foliations $\mathcal{F}, \tilde{\mathcal{F}}$, which are $\mathcal{C}^1$-close, we say that $\mathcal{F}$ intersects transversally $\tilde{\mathcal{F}}$—denoted as $\mathcal{F} \pitchfork \tilde{\mathcal{F}}$—when given any leaf of $\mathcal{F}$, we can find another leaf of $\tilde{\mathcal{F}}$ for which there is a non-trivial intersection which is transversal. (There could be other non-transversal intersections.)

Note that our use of "foliation intersects transversely "is at variance with standard use in differential topology where it is often taken to mean that, given a leaf of $\mathcal{F}$ and a leaf of $\tilde{\mathcal{F}}$, they either intersect transversally or do not intersect.

To show the transversality between the foliations $\mathcal{F}_F$ and $\mathcal{F}_{F \circ S^{-1}}$ we only need to obtain lower bounds for the angle between the parameterized surface

$$S(L_E^F) = \{S(\lambda_E(\varphi, s; \varepsilon), \varphi, s)\}$$

and the implicit surface

$$L_{E'}^F = \{(I, \varphi, s), \ F(I, \varphi, s; \varepsilon) = E'\}.$$

More precisely, we will compute $\sin(\alpha)$, where $\alpha$ is the angle between the tangent planes to $L_{E'}^F$ and $S(L_E^F)$.

Recalling that the normal vector to the tangent plane to $L_{E'}^F$ is given by $\frac{\nabla_{I,\varphi,s}F}{|\nabla_{I,\varphi,s}F|}$, and that any vector of the tangent plane to $S(L_E^F)$ is written as $D_v(S \circ (\lambda_E, \mathrm{Id}, \mathrm{Id})) = D(S \circ (\lambda_E, \mathrm{Id}, \mathrm{Id}))v$, we obtain:

$$(10.2) \qquad \sin(\alpha) = \max_{v \in \mathbb{R}^2} \frac{|D_v F \circ S \circ (\lambda_E, \mathrm{Id}, \mathrm{Id})|}{|(\nabla_{I,\varphi,s}F) \circ (\lambda_E, \mathrm{Id}, \mathrm{Id})| \, |D_v S \circ (\lambda_E, \mathrm{Id}, \mathrm{Id})|}.$$

Hence, we will obtain lower bounds for the angles taking vectors $v$ that make the computations in the right hand side of (10.2) simple.

LEMMA 10.7. *Let*

$$F : \mathcal{A} = (I_1, I_2) \times \mathcal{J} \times (-\varepsilon_0, \varepsilon_0) \to \mathbb{R}$$

*be a $\mathcal{C}^r$ function, $r \geq 2$, where $\mathcal{J} \subset \mathbb{T}^2$ is an open set.*

*Assume that for any $(I, \varphi, s) \in (I_1, I_2) \times \mathcal{J}$, the equation $F(I, \varphi, s; \varepsilon) = E$, for $E \in (E_1, E_2) = F(\mathcal{A})$ defines a smooth surface $L_E^F$ given as a graph*

$$I = \lambda_E(\varphi, s; \varepsilon).$$

*Let $S$ be the scattering map which has been computed to first order in (9.9) and assume that there exists an open set $\mathcal{J}' \subset \mathcal{J}$, $\mathcal{J}' \neq \emptyset$, and a constant $C > 0$ independent of $\varepsilon$ and $E$, such that for any $(\varphi, s) \in \mathcal{J}'$ one has,*

$$(10.3) \qquad \frac{|\nabla_{\varphi,s}(F \circ S \circ (\lambda_E, \mathrm{Id}, \mathrm{Id})(\varphi, s))|}{|\nabla_{I,\varphi,s}F(\lambda_E(\varphi, s; \varepsilon), \varphi, s)|} \geq C\varepsilon.$$

*Then, if we denote by $\mathcal{F}_F$ the foliation given by*

$$\{(I, \varphi, s), \; F(I, \varphi, s; \varepsilon) = E, \; E \in (E_1, E_2)\} = \bigcup_{E \in (E_1, E_2)} L_E^F,$$

*we have:*

$$\mathcal{F}_F \pitchfork \mathcal{F}_{F \circ S^{-1}}.$$

*More precisely, there exists a constant $C'$, independent of $\varepsilon$ and $E$, such that the angle between the surfaces $S(L_E^F)$ and $L_{E'}^F$ at the intersection can be bounded from below by $C'\varepsilon$.*

PROOF. To show the transversality between the foliations $\mathcal{F}_F$ and $\mathcal{F}_{F \circ S^{-1}}$, we only need to obtain lower bounds for the angle (10.2) between the parameterized surface $S(\lambda_E(\varphi, s; \varepsilon), \varphi, s)$ and the implicit surface $F(I, \varphi, s; \varepsilon) = E'$.

By formula (9.9), we have $S = \mathrm{Id} + \varepsilon S_1 + \mathrm{O}_{\mathcal{C}^1}(\varepsilon^{1+\varrho})$. Hence, there exists a constant $\bar{C}$, independent of $\varepsilon$ such that

$$|D_v S \circ (\lambda_E, \mathrm{Id}, \mathrm{Id})| \leq \bar{C} |v|$$

and formula (10.2) is bounded if (10.3) is verified. Then, in order to obtain that the foliations intersect transversally we only need to assume condition (10.3), and we obtain that the angle of intersection is bounded from below by $C'\varepsilon$, where $C'$ is some suitable constant. □

**10.2.1. The non-resonant region and resonances of order $3$ and higher.** Now, we apply Lemma 10.7 to the non-resonant region $\mathcal{S}^L$ and the resonant regions $\mathcal{S}^{\mathcal{R}_j}$, for $j \geq 3$, where the tori, given in Propositions 8.21 and 8.24—see also Remarks 8.22 and 8.25—, are rather "flat". That is, we can take $F(I, \varphi, s; \varepsilon) = I + \mathrm{O}_{\mathcal{C}^2}(\varepsilon)$ in the previous Lemma 10.7, and $\lambda_E(\varphi, s; \varepsilon) = E + U_E(\varphi, s; \varepsilon)$, where $|U_E|_{\mathcal{C}^2} \leq \mathrm{cte.}\,\varepsilon$ as given in Remarks 8.22 and 8.25.

LEMMA 10.8. *Consider a foliation $\mathcal{F}_F$, contained in a connected component of the non resonant region $\mathcal{S}^L$ defined in (8.15) (or of the resonant regions $\mathcal{S}^{\mathcal{R}_j}$, defined in (8.28), for $j \geq 3$).*

*Assume that the function $F$ is of the form $F(I, \varphi, s; \varepsilon) = I + \mathrm{O}_{\mathcal{C}^2}(\varepsilon)$, so that equation $F(I, \varphi, s; \varepsilon) = E$ defines a smooth surface given as a graph $I = \lambda_E(\varphi, s; \varepsilon) = E + \mathrm{O}_{\mathcal{C}^2}(\varepsilon)$, for $(\varphi, s) \in \mathbb{T}^2$.*

*Assume also that the second part of Hypothesis **H4** is fulfilled, more precisely, that the reduced Poincaré function $\mathcal{L}^*$ defined in (9.7) verifies, for any value of $(I, \varphi, s) \in H_- \cap \mathcal{S}^L$ (respectively for $(I, \varphi, s) \in H_- \cap \mathcal{S}^{\mathcal{R}_j}$) that the function*

$$(10.4) \qquad \theta \mapsto \frac{\partial \mathcal{L}^*}{\partial \theta}(E, \theta)$$

*is negative and non-constant for $\theta \in \mathcal{J}_E^*$ (see (9.8)).*

*Then, the foliations $\mathcal{F}_F$ and $\mathcal{F}_{F \circ S^{-1}}$ intersect transversally.*

*More precisely, there exist constants, $0 < C, C', C''$, independent of $\varepsilon$ and $E$, such that any surface $S(L_E^F)$ intersects at some point the surface $L_{E'}^F$ for any $E'$, $C\varepsilon < E' - E \leq C''\varepsilon$.*

*The angle between the surfaces $S(L_E^F)$ and $L_{E'}^F$ at the intersection can be bounded from below by $C'\varepsilon$.*

REMARK 10.9. We know, by Propositions 8.21 and 8.24, that the gaps between two consecutive tori are, at most, of order $\bar{I} - I \simeq \varepsilon^{3/2}$ and that the tori are $O_{\mathcal{C}^1}(\varepsilon)$ close to the level sets of the action $I$.

Then, when we apply Lemma 10.8 to these tori, and thanks to the non-constantness and negativity of the map (4.6), we obtain that the image under the scattering map of a torus in this region given by $I = I_0 + O(\varepsilon)$, intersects transversally any other torus given by $I = \bar{I}_0 + O(\varepsilon)$ with $\bar{A}\varepsilon \leq \bar{I}_0 - I_0 \leq A\varepsilon$.

PROOF. We apply Lemma 10.7, with $F(I, \varphi, s; \varepsilon) = I + O_{\mathcal{C}^2}(\varepsilon)$, and $\lambda_E(\varphi, s; \varepsilon) = E + O_{\mathcal{C}^2}(\varepsilon)$. To check inequality (10.3) we use formula (9.9), so that $S = \mathrm{Id} + \varepsilon S_1 + O_{\mathcal{C}^1}(\varepsilon^{1+\varrho})$. Hence, we compute:

$$
\begin{aligned}
F \circ S &\circ (\lambda_E, \mathrm{Id}, \mathrm{Id})(\varphi, s) \\
&= F(\lambda_E(\varphi, s; \varepsilon), \varphi, s; \varepsilon) \\
&\quad + \varepsilon \nabla_{I, \varphi, s} F(\lambda_E(\varphi, s; \varepsilon), \varphi, s; \varepsilon) \left(S_1(\lambda_E(\varphi, s; \varepsilon), \varphi, s; \varepsilon) + O_{\mathcal{C}^1}(\varepsilon^{\varrho})\right) \\
&\quad + O_{\mathcal{C}^1}(\varepsilon^2) \\
&= E - \varepsilon \frac{\partial \mathcal{L}^*}{\partial \theta}(\lambda_E(\varphi, s; \varepsilon), \varphi - \lambda_E(\varphi, s; \varepsilon)s) + O_{\mathcal{C}^1}(\varepsilon^{1+\varrho}) \\
&= E - \varepsilon \frac{\partial \mathcal{L}^*}{\partial \theta}(E, \varphi - Es) + O_{\mathcal{C}^1}(\varepsilon^{1+\varrho}),
\end{aligned}
$$

and $|\nabla_{I, \varphi, s} F(\lambda_E(\varphi, s; \varepsilon), \varphi, s; \varepsilon)| = 1 + O_{\mathcal{C}^1}(\varepsilon)$. If for any value of $(E, \theta)$, the function $\frac{\partial \mathcal{L}^*}{\partial \theta}(E, \theta)$ is non-constant, there exists an interval $\bar{\mathcal{J}}_E \subset \mathcal{J}_E^*$ where

$$
\left| \frac{\partial^2 \mathcal{L}^*}{\partial \theta^2} \right| \geq C > 0.
$$

Then, hypothesis (10.3) is verified in $\bar{\mathcal{J}}_E$ and Lemma 10.7 applies. On the other hand, as $\frac{\partial \mathcal{L}^*}{\partial \theta}(E, \theta) < 0$, the surface $S(L_E^F)$ intersect surfaces $L_{E'}^F$, for $E' > E$, and $E' - E = O(\varepsilon)$. $\square$

**10.2.2. Resonances of first order.** Now we turn to the task of identifying lower bounds for the angles of intersection. The main result will be Lemma 10.11

To obtain more manageable final conditions, we will use the following technical Lemma:

LEMMA 10.10. *Let $a(\theta)$, $b(\theta)$ be functions of class $\mathcal{C}^r$, $r \geq 0$, such that the function $a(\theta)/b(\theta)$ is not constant in some interval $\mathcal{J}$.*

*Then, there exists a constant $\tilde{C} > 0$ and two intervals $\mathcal{J}_1$, $\mathcal{J}_2$ subsets of $\mathcal{J}$ such that, given any value $(x, y) \in \mathbb{R}^2$, one can choose $\theta$ which belongs to one of the two intervals $\mathcal{J}_1$ $\mathcal{J}_2$ and such that:*

$$(10.5) \qquad |a(\theta)x + b(\theta)y| \geq \tilde{C}(|x| + |y|)$$

PROOF. If $a(\theta)/b(\theta)$ is not constant in $\mathcal{J}$ there exist at least two values $\theta_i \in \mathcal{J}$, $i = 1, 2$ (and intervals $\mathcal{J}_i$ around them) such that the matrix

$$A = \left( \begin{array}{cc} a(\theta_1) & b(\theta_1) \\ a(\theta_2) & b(\theta_2) \end{array} \right)$$

is invertible. Then, given any vector $z = (x, y)$, we have that

$$|Az| \geq \frac{|z|}{|A^{-1}|}$$

where $|\cdot|$ stands for the sup norm. Given $z$, we choose $\theta_i$, for $i = 1$ or $i = 2$, such that $|Az| = |a(\theta_i)x + b(\theta_i)y|$, and taking into account that we are dealing with continuous functions we obtain that:

$$|a(\theta)x + b(\theta)y| \geq \frac{1}{|A^{-1}|} |z|$$

in $\mathcal{J}_i$. The proof finishes taking $\tilde{C} < \frac{1}{|A^{-1}|}$. $\qquad\qquad\square$

¿From now on, we will apply Lemma 10.7 to study the resonant regions $\mathcal{S}^{\mathcal{R}_j}$, $j = 1, 2$. In these regions, the tori and the stable and unstable manifolds of the periodic orbit, are not flat as we showed in Theorem 8.30 and Proposition 8.40. The fact that the tori are not flat has the consequence that the dominant effect of comparing a torus with the torus in the image of the scattering map will include some extra terms.

We recall that in Theorem 8.30 we showed that the invariant objects in the resonant regions $\mathcal{S}^{\mathcal{R}_j}$, $j = 1, 2$ are given very approximately by the level sets of the Hamiltonian $K_0(y, x; \varepsilon)$ in (8.41), when written in the averaged variables $(y, x, s)$.

The relation between these variables and the original ones $(I, \varphi, s)$ are the changes of variables given in Proposition 8.2, Theorem 8.9 and in (8.32), (8.35), and is given in first order by

$$y = k_0 I + l_0 + O_{\mathcal{C}^2}(\varepsilon).$$

Then, using (8.39), we obtain that all these objects are given by the level sets of a function:

$$(10.6) \qquad F(I, \varphi, s; \varepsilon) = \frac{(k_0 I + l_0 + O_{\mathcal{C}^2}(\varepsilon))^2}{2}(1 + O_{\mathcal{C}^2}(\varepsilon)) + O_{\mathcal{C}^2}(\varepsilon^j),$$

where $j = 1, 2$ is the order of the resonance.

Moreover, in Corollary 8.31 and in Proposition 8.40 the KAM tori (primary and secondary) and the (un)stable manifolds of periodic orbits were written as graphs in the variables $(y, x, s)$ of functions of the form

$$y = \mathcal{Y}_\pm(x, E) + O_{\mathcal{C}^1}(\varepsilon^{3/2}).$$

In remarks 8.32 and 8.41 we have obtained the corresponding expressions in the original variables $(I, \varphi, s)$. The dominant terms in $F$ and in the expressions of these invariant objects will be different whether the resonance is of order 1 or 2. The hypothesis of the Lemmas 10.11 and 10.14 are tailored to apply to resonances of order 1 or 2.

LEMMA 10.11. *Let us consider a foliation $\mathcal{F}_F$ in a connected component of the resonant region $\mathcal{S}^{\mathcal{R}_1}$, defined in (8.28), where $\mathcal{R}_1$ is given in (5.4). More precisely, consider a resonance $-l_0/k_0 \in \mathcal{R}_1$, and in the region*

$$\{(I, \varphi, s) \in [l_0/k_0 - L, l_0/k_0 + L] \times \mathbb{T}^2\},$$

*the function $F$ is of the form*

$$F(I, \varphi, s; \varepsilon) = \frac{(k_0 I + l_0)^2}{2} + O_{\mathcal{C}^2}(\varepsilon),$$

*and, for some $0 < \rho < \pi$, and for some range of energies $-c_4 \varepsilon \leq E \leq c_2 L$, the equation $F(I, \varphi, s; \varepsilon) = E$, defines two smooth surfaces $S(L_E^{F,\pm})$ given as a graph $I = \lambda_E^{\pm}(\varphi, s; \varepsilon)$ with:*

$$(10.7) \quad \begin{aligned} \lambda_E^{\pm}(\varphi, s; \varepsilon) &= -\frac{l_0}{k_0} + \frac{1}{k_0} \mathcal{Y}_{\pm}(\theta, E) + O_{\mathcal{C}^2}(\varepsilon) \\ &= -\frac{l_0}{k_0} \pm \frac{1}{k_0}(1 + \varepsilon b)\ell(\theta, E) + \frac{\varepsilon}{k_0} \tilde{\mathcal{Y}}_{\pm}(\ell(\theta, E)) + O_{\mathcal{C}^2}(\varepsilon), \end{aligned}$$

*for $\rho \leq \theta := k_0 \varphi + l_0 s \leq 2\pi - \rho$, where $\ell(\theta, E) = \sqrt{2(E - \varepsilon U(\theta; \varepsilon))}$ with $U(\theta; \varepsilon)$ defined in (8.37) and $\tilde{\mathcal{Y}}_{\pm}$ is given in Lemma 8.34, with $\delta = \varepsilon$.*

*Given the reduced Poincaré function $\mathcal{L}^*$, defined in (9.7), for $(I, \varphi, s) \in H_- \cap \mathcal{S}^{\mathcal{R}_1}$, let us assume that the second part of Hypothesis **H4** is fulfilled, more precisely, that the function*

$$(10.8) \quad \theta \mapsto \frac{\partial \mathcal{L}^*}{\partial \theta}(I, \theta)$$

*is non-constant and negative for $\theta \in \mathcal{J}_I^*$ (see (9.8)). Assume, moreover, the following hypothesis, which is **H5"** in Theorem 4.1.*

*The function*

$$(10.9) \quad \theta \mapsto \frac{k_0 U'(\theta; 0) \frac{\partial \mathcal{L}^*}{\partial \theta}(-\frac{l_0}{k_0}, \frac{\theta}{k_0}) + 2U(\theta; 0) \frac{\partial^2 \mathcal{L}^*}{\partial \theta^2}(-\frac{l_0}{k_0}, \frac{\theta}{k_0})}{2 \frac{\partial^2 \mathcal{L}^*}{\partial \theta^2}(-\frac{l_0}{k_0}, \frac{\theta}{k_0})}$$

*is non-constant.*

*Then, the foliations $\mathcal{F}_F$ and $\mathcal{F}_{F \circ S^{-1}}$ intersect transversally.*

*More precisely, there exist constants, $0 < C, C', C''$, independent of $\varepsilon$ and $E$, such that:*

    (1) *Any surface $S(L_E^{F,-})$ intersects at some point the surface $L_{E'}^{F,-}$ for any $E'$ such that*

$$C\varepsilon \max\left(|E|^{1/2}, \varepsilon^{1/2}\right) \leq E - E' \leq C''\varepsilon \max\left(|E|^{1/2}, \varepsilon^{1/2}\right)$$

(2) *Any surface $S(L_E^{F,+})$ intersects at some point the surface $L_{E'}^{F,+}$ for any $E'$ such that*

$$C\varepsilon \max\left(|E|^{1/2}, \varepsilon^{1/2}\right) \le E' - E \le C''\varepsilon \max\left(|E|^{1/2}, \varepsilon^{1/2}\right)$$

*The angle between the surfaces $S(L_E^{F,\pm})$ and $L_{E'}^{F,\pm}$ at the intersection is bounded from below by $C'\varepsilon$.*

REMARK 10.12. We know, by Theorem 8.30 that all the tori in the resonant region are given by formulas (10.7) for $E = E_i, F_i, G$ and verify

$$\begin{aligned}
|E_i - E_{i+1}| &\le \varepsilon^{\frac{3}{2}+\frac{1}{2}} \le \max(\sqrt{E_i}, \varepsilon^{1/2}), \\
|E_{l_E} - F_1| &\le \varepsilon^{\frac{3}{2}+\frac{1}{2}} \le \max(\sqrt{E_{l_E}}, \varepsilon^{1/2}), \\
|F_i - F_{i+1}| &\le \varepsilon^{\frac{3}{2}+\frac{1}{2}} \le \max(\sqrt{F_i}, \varepsilon^{1/2}) \\
|F_{l_F} - G| &\le \varepsilon^{\frac{3}{2}+\frac{1}{2}} \le \max(\sqrt{F_{l_F}}, \varepsilon^{1/2}).
\end{aligned}$$

and that by and Corollary 8.31 and Remark 8.32 that, when they are written as graphs, the distance between to consecutive tori is of order $O(\varepsilon^{3/2})$.

Then, when we apply Lemma 10.11 to these tori we obtain that the image under the scattering map of a torus in this region intersect transversally another torus.

PROOF. It suffices to apply Lemma 10.7 to the function

$$F(I, \varphi, s; \varepsilon) = \frac{(k_0 I + l_0)^2}{2} + O_{\mathcal{C}^2}(\varepsilon),$$

and $\lambda_E^{\pm}$ given by (10.7).

Note that to study transversality of the foliations, we can consider $E$ fixed since $E$ is only a label for the leaves.

Since we consider $E$ fixed, the analysis of dominant terms will only involve estimating derivatives with respect to the angle variables.

We will also see that the angle variables will enter in the dominant terms only trough $\theta = k_0\varphi + l_0 s$ (the resonant angle).

In order to check inequality (10.3) we use formula (9.9), so that $S = \mathrm{Id} + \varepsilon S_1 + O_{\mathcal{C}^1}(\varepsilon^{1+\varrho})$, and that, by (8.57), $\lambda_E^{\pm}$ is a bounded function, and its derivatives with respect to $\varphi, s$ are of order $O(\varepsilon^{1/2})$. Hence, we can compute:

$$
\begin{aligned}
&F \circ So(\lambda_E^\pm, \mathrm{Id}, \mathrm{Id})(\varphi, s) \\
&= F(\lambda_E^\pm(\varphi, s; \varepsilon), \varphi, s; \varepsilon) + \varepsilon \nabla_{I,\varphi,s} F(\lambda_E^\pm(\varphi, s; \varepsilon), \varphi, s; \varepsilon) \cdot \\
&\quad \left( S_1(\lambda_E^\pm(\varphi, s; \varepsilon), \varphi, s) + O_{\mathcal{C}^1}(\varepsilon^\varrho) \right) + O_{\mathcal{C}^1}(\varepsilon^2) \\
&= E - \varepsilon k_0 (k_0 \lambda_E^\pm(\varphi, s; \varepsilon) + l_0) \cdot \\
&\quad \left( \frac{\partial \mathcal{L}^*}{\partial \theta}(\lambda_E^\pm(\varphi, s; \varepsilon), \varphi - \lambda_E^\pm(\varphi, s; \varepsilon)s) + O_{\mathcal{C}^1}(\varepsilon^\varrho) \right) + O_{\mathcal{C}^1}(\varepsilon^2) \\
&= E - \varepsilon k_0 \mathcal{Y}_\pm(\theta, E) \cdot \\
&\quad \left( \frac{\partial \mathcal{L}^*}{\partial \theta}\left( -\frac{l_0}{k_0} + \frac{1}{k_0}\mathcal{Y}_\pm(\theta, E), \varphi - \left( -\frac{l_0}{k_0} + \frac{1}{k_0}\mathcal{Y}_\pm(\theta, E) \right) s \right) + O_{\mathcal{C}^1}(\varepsilon^\varrho) \right) \\
&\quad + O_{\mathcal{C}^1}(\varepsilon^2) \\
&= E \mp \varepsilon k_0 (1 + \varepsilon b) \ell(\theta, E) \cdot \\
&\quad \left( \frac{\partial \mathcal{L}^*}{\partial \theta}\left( -\frac{l_0}{k_0} \pm \frac{(1 + \varepsilon b)}{k_0}\ell(\theta, E), \varphi - \left( -\frac{l_0}{k_0} \pm \frac{(1 + \varepsilon b)}{k_0}\ell(\theta, E) \right) s \right) \right. \\
&\quad \left. + O_{\mathcal{C}^1}(\varepsilon^\varrho) \right) + O_{\mathcal{C}^1}(\varepsilon^2) \\
&= E \mp \varepsilon \mathcal{M}(\theta; \varepsilon) + O_{\mathcal{C}^1}(\varepsilon^{1+\varrho})
\end{aligned}
$$

Now, we compute the main part of the function $\mathcal{M}$

$$
\begin{aligned}
\mathcal{M}(\theta; \varepsilon) = k_0 \ell(\theta, E) \cdot \\
\frac{\partial \mathcal{L}^*}{\partial \theta}\left( -\frac{l_0}{k_0} + \frac{1}{k_0}\ell(\theta, E), \varphi - \left( -\frac{l_0}{k_0} + \frac{1}{k_0}\ell(\theta, E) \right) s \right)
\end{aligned}
$$

The analysis of the function $\mathcal{M}$ will be done differently whether we are close to the resonance or whether we are reasonably far because in these cases the dominant terms will be different. We choose any $\nu$ such that $0 \leq \nu < 1$, and we will consider the following two cases:

**Close to the resonance:** $-c_4 \varepsilon \leq E \leq \tilde{c} \varepsilon^\nu$.
     In this region, we have that $|\ell(\cdot, E)|_{\mathcal{C}^1} \leq \varepsilon^{\nu/2}$, and then, by bounds (8.57), the dominant terms in $\mathcal{M}$ are

$$
\begin{aligned}
\mathcal{M}(\theta; \varepsilon) = \\
= k_0 \sqrt{2(E - \varepsilon U(\theta; 0))} \left( \frac{\partial \mathcal{L}^*}{\partial \theta}\left( -\frac{l_0}{k_0}, \frac{\theta}{k_0} \right) + O_{\mathcal{C}^1}(\varepsilon^{\nu/2}) \right).
\end{aligned}
$$

To apply Lemma 10.7 we need to check (10.3), that is, it suffices to show that we can bound from below the derivative of this function divided by $k_0\lambda_E^\pm(\varphi, s; \varepsilon) + l_0$. Computing this derivative, we obtain:

$$\frac{\partial}{\partial\theta}\left(\mathcal{M}(\theta;\varepsilon)\right) =$$

$$= \frac{k_0}{\sqrt{2(E - \varepsilon U(\theta; 0))}}\left(2E\frac{\partial^2}{\partial\theta^2}\mathcal{L}^*(-\frac{l_0}{k_0}, \frac{\theta}{k_0})\right.$$

$$\left. -\varepsilon\left[k_0 U'(\theta; 0)\frac{\partial\mathcal{L}^*}{\partial\theta}(-\frac{l_0}{k_0}, \frac{\theta}{k_0}) + 2U(\theta; 0)\frac{\partial^2}{\partial\theta^2}\mathcal{L}^*(-\frac{l_0}{k_0}, \frac{\theta}{k_0})\right]\right)$$

$$+ \sqrt{2(E - \varepsilon U(\theta; 0))}\, O_{\mathcal{C}^0}(\varepsilon^{\nu/2}) + O_{\mathcal{C}^0}(\varepsilon^{(1+\nu)/2}),$$

and then,

$$\frac{\frac{\partial}{\partial\theta}\left(\mathcal{M}(\theta;\varepsilon)\right)}{k_0\lambda_E^\pm(\varphi, s; \varepsilon) + l_0} =$$

$$\frac{\pm k_0}{2(E - \varepsilon U(\theta; 0))}\left(2E\frac{\partial^2}{\partial\theta^2}\mathcal{L}^*(-\frac{l_0}{k_0}, \frac{\theta}{k_0})\right.$$

$$\left. -\varepsilon\left[k_0 U'(\theta; 0)\frac{\partial\mathcal{L}^*}{\partial\theta}(-\frac{l_0}{k_0}, \frac{\theta}{k_0}) + 2U(\theta; 0)\frac{\partial^2}{\partial\theta^2}\mathcal{L}^*(-\frac{l_0}{k_0}, \frac{\theta}{k_0})\right]\right)$$

$$+ O_{\mathcal{C}^0}(\varepsilon^{\nu/2}),$$

Now, we apply Lemma 10.10, with $x = -\varepsilon$, $y = E$, and

$$a(\theta) = k_0 U'(\theta; 0)\frac{\partial\mathcal{L}^*}{\partial\theta}(-\frac{l_0}{k_0}, \frac{\theta}{k_0}) + 2U(\theta; 0)\frac{\partial^2\mathcal{L}^*}{\partial\theta^2}(-\frac{l_0}{k_0}, \frac{\theta}{k_0})$$

$$b(\theta) = 2\frac{\partial^2\mathcal{L}^*}{\partial\theta^2}(-\frac{l_0}{k_0}, \frac{\theta}{k_0}).$$

Using Lemma 10.10, and that

$$|2(E - \varepsilon U(\theta, 0))| \leq \text{cte.}\,(|E| + |\varepsilon|),$$

we obtain that for $\theta$ either in $\mathcal{J}_1$, $\mathcal{J}_2$, where $\mathcal{J}_i \subset \mathcal{J}_{-l_0/k_0}^*$, $i = 1, 2$

$$\left|\frac{\frac{\partial}{\partial\theta}\left(\mathcal{M}(\theta;\varepsilon)\right)}{k_0\lambda_E^\pm(\varphi, s; \varepsilon) + l_0}\right| \geq \tilde{C}.$$

Consequently, the angle of intersection can be bounded from below by $C'\varepsilon$, for some suitable constant independent of $\varepsilon$.

To identify more precisely which surfaces will intersect $S(L_E^{F,\pm})$, we only need to observe that the function $\mathcal{M}(\theta; \varepsilon)$ is given approximately by

$$k_0\sqrt{2(E - \varepsilon U(\theta; 0))}\frac{\partial\mathcal{L}^*}{\partial\theta}(-\frac{l_0}{k_0}, \frac{\theta}{k_0}),$$

and then it is a negative function. On the other hand we have that:

$$F \circ S \circ (\lambda_E^\pm, \text{Id}, \text{Id})(\varphi, s) \simeq E \mp \varepsilon\mathcal{M}(\theta; \varepsilon),$$

then the surface $S(L_E^{F,-})$ intersect surfaces $L_{E'}^{F,-}$ with $E' < E$. (Note that if $E' < E$ then $\lambda_E^-(\varphi, s; \varepsilon) < \lambda_{E'}^-(\varphi, s; \varepsilon)$). Moreover, one has that there exists some constant $C''$ such that

$$\max_{\theta \in \mathcal{J}_1 \cup \mathcal{J}_2} |\mathcal{M}(\varphi, s; \varepsilon)| \geq C'' \max (E^{1/2}, \varepsilon^{1/2}),$$

and applying Lemma 10.3 we obtain the desired result.

An analogous result is obtained for $\lambda_E^+$, with $E' > E$.

**Far from the resonance:** $\tilde{c}\varepsilon^\nu \leq E \leq c_2 L$.

This case is analogous to the non resonant region, because in this case,

$$\begin{aligned}
\ell(\theta, s) &= \sqrt{2(E - \varepsilon U(\theta; \varepsilon))} = \sqrt{2E} \sqrt{1 - \frac{\varepsilon}{E} U(\theta; \varepsilon)} \\
&= \sqrt{2E}(1 + \mathrm{O}_{\mathcal{C}^2}(\varepsilon^{1-\nu})),
\end{aligned}$$

consequently, the function $\mathcal{M}$ becomes

$$\begin{aligned}
&\mathcal{M}(\theta; \varepsilon) \\
&= k_0 \sqrt{2E} \frac{\partial \mathcal{L}^*}{\partial \theta}\left(-\frac{l_0}{k_0} + \sqrt{2E}, \varphi - (-\frac{l_0}{k_0} + \sqrt{2E})s\right) + \mathrm{O}_{\mathcal{C}^2}(\varepsilon^{1-\nu}).
\end{aligned}$$

Then, if the function $\mathcal{L}^*(I, \theta)$ is not constant, we apply Lemma 10.7, to get the desired result.

$\square$

REMARK 10.13. We observe that the transversality lemmas 10.8 and 10.11 use some non-degeneracy hypothesis. Hypothesis (10.4) and (10.8) refer to the reduced Poincaré function $\mathcal{L}^*$, and therefore to $\mathcal{L}$ (see (9.7)), and both are contained in the second part of Hypothesis **H4** of Theorem 4.1. The hypothesis that the function (10.9) is non-constant also involves the function $U(\theta; \varepsilon)$, which comes from the expression of the Hamiltonian reduced to $\tilde{\Lambda}_\varepsilon$ near the resonance $I = -\frac{l_0}{k_0}$. This Hypothesis is now explicitly stated in Lemma 10.11, (10.9). This is **H5"** in Theorem 4.1.

**10.2.3. Resonances of order 2.** Now we turn to study resonances of order 2. These are the hardest to study since the size of the resonant region is $\mathrm{O}(\varepsilon)$, which is the same order of magnitude than the scattering map. This causes that there are different geometries that could happen, depending on the numerical values of the leading coefficients. If the size of change induced by the scattering map, is larger than the size of the resonant region, it could happen that the scattering map connects primary KAM tori on one side of the resonance to primary tori on the other side of the resonance. On the other hand, if the size of the resonance is smaller than the change induced by the scattering map, we will need to use the secondary tori. These two alternatives will be identified by remembering that, according to the second part of Hypothesis **H4**, the $I$ variable always

increases and keeping track of the values of the averaged energy in after the
scattering map is applied. Positive energy corresponds to primary tori and
negative energy to secondary.

LEMMA 10.14. *Consider a foliation $\mathcal{F}_F$ in a connected component of the
resonant region $\mathcal{S}^{\mathcal{R}_2}$, defined in (8.28), where $\mathcal{R}_2$ is given in (5.5). More
precisely, consider a resonance $-l_0/k_0 \in \mathcal{R}_2$, and in the region*

$$\{(I, \varphi, s) \in [l_0/k_0 - L, l_0/k_0 + L] \times \mathbb{T}^2\},$$

*the function $F$ is of the form*

$$F(I, \varphi, s; \varepsilon) = \frac{y^2}{2}(1 + O_{\mathcal{C}^2}(\varepsilon)) + O_{\mathcal{C}^2}(\varepsilon^2),$$

*where $y = y(I, \varphi, s; \varepsilon)$ is the variable defined in this resonant region through
the changes of variables given in Proposition 8.2, Theorem 8.9 and in 8.32,
8.35, given, up to first order, by*

(10.10)
$$\begin{aligned}
y &= k_0 I + l_0 + O_{\mathcal{C}^2}(\varepsilon), \\
x &= \theta + O_{\mathcal{C}^2}(\varepsilon), \quad \theta = k_0 \varphi + l_0 s.
\end{aligned}$$

*For some $0 < \rho < \pi$, and for some range of energies $-c_4 \varepsilon^2 \leq E \leq c_2 L$,
the equation $F(I, \varphi, s; \varepsilon) = E$, defines two smooth surfaces $S(L_E^{F,\pm})$ given as
graphs $I = \lambda_E^{\pm}(\varphi, s; \varepsilon)$. These functions $\lambda_E^{\pm}$ are related by the change (10.10)
with $f_E^{\pm}(x, s; \varepsilon)$, which are the graphs of the surfaces in the variables $(y, x, s)$:*

(10.11)
$$\lambda_E^{\pm}(\varphi, s; \varepsilon) = -\frac{l_0}{k_0} + \frac{1}{k_0} f_E^{\pm}(x, s; \varepsilon) + O_{\mathcal{C}^2}(\varepsilon),$$

*and*

(10.12)
$$\begin{aligned}
f_E^{\pm}(x, s; \varepsilon) &= \mathcal{Y}_{\pm}(x, E) + O_{\mathcal{C}^2}(\varepsilon^{3/2}) \\
&= \pm(1 + \varepsilon b)\ell(x, E) + \varepsilon \tilde{\mathcal{Y}}_{\pm}(\ell(x, E)) + O_{\mathcal{C}^2}(\varepsilon^{3/2}),
\end{aligned}$$

*for $\rho \leq \theta \leq 2\pi - \rho$, where $\ell(x, E) = \sqrt{2(E - \varepsilon^2 U(x; \varepsilon))}$ with $U(x; \varepsilon)$ defined
in (8.37) and $\tilde{\mathcal{Y}}_{\pm}$ is given in Lemma 8.34 with $\delta = \varepsilon^2$.*

*Given the reduced Poincaré function $\mathcal{L}^*$, defined in (9.7), for $(I, \varphi, s) \in
H_- \cap \mathcal{S}^{\mathcal{R}_2}$, let us assume that the second part of Hypothesis **H4** is fulfilled,
more precisely, that the the function*

(10.13)
$$\theta \mapsto \frac{\partial \mathcal{L}^*}{\partial \theta}(I, \theta)$$

*is non-constant and negative for $\theta \in \mathcal{J}_I^*$ (see (9.8)).*

*Assume moreover the following hypothesis, which is **H5"** in Theorem 4.1.*

*There exists a constant $C$, independent of $E$ and $\varepsilon$, and an interval
$\mathcal{J} \subset \mathcal{J}_{-l_0/k_0}^*$, such that, given any $E, \varepsilon$ in this region $(-c_4 \varepsilon^2 \leq E \leq c_2 L)$*

*and* $\theta \in \mathcal{J}$,

$$
(10.14) \quad \left| \frac{k_0}{2(E - \varepsilon^2 U(\theta; 0))} \left( 2E \frac{\partial^2 \mathcal{L}^*}{\partial \theta^2} (-\frac{l_0}{k_0}, \frac{\theta}{k_0}) \right. \right.
$$

$$
- \varepsilon^2 \left[ k_0 U'(\theta, 0) \frac{\partial \mathcal{L}^*}{\partial \theta} (-\frac{l_0}{k_0}, \frac{\theta}{k_0}) + 2U(\theta; 0) \frac{\partial^2 \mathcal{L}^*}{\partial \theta^2} (-\frac{l_0}{k_0}, \frac{\theta}{k_0}) \right]
$$

$$
\left. \left. \pm \varepsilon \sqrt{2(E - \varepsilon^2 U(\theta; 0))} \frac{\partial^2 \mathcal{L}^*}{\partial \theta^2} (-\frac{l_0}{k_0}, \frac{\theta}{k_0}) \right) \right| \geq C
$$

Then, the foliations $\mathcal{F}_F$ and $\mathcal{F}_{F \circ S^{-1}}$ intersect transversally.

More precisely, there exist constants, $C$, $C'$, $C''$, independent of $\varepsilon$ and $E$, such that:

(1) Given a surface $S(L_E^{F,-})$ we have one of the following cases:
  (a) $S(L_E^{F,-})$ intersects at some point the surface $L_{E'}^{F,-}$, for any $E$ such that $C\varepsilon \max(|E|^{1/2}, \varepsilon) \leq E - E' \leq C''\varepsilon \max(|E|^{1/2}, \varepsilon)$,
  (b) $S(L_E^{F,-})$ intersects at some point the surface $L_{E'}^{F,+}$, for any $E'$ such that $C\varepsilon \max(|E|^{1/2}, \varepsilon) \leq E' - E \leq C''\varepsilon \max(|E|^{1/2}, \varepsilon)$.
(2) Any surface $S(L_E^{F,+})$ intersects at some point the surface $L_{E'}^{F,+}$, for any $E$ such that $C\varepsilon \max(|E|^{1/2}, \varepsilon) \leq E' - E \leq C''\varepsilon \max(|E|^{1/2}, \varepsilon)$.

In all the cases the angle between these surfaces at the intersection can be bounded from below by $C'\varepsilon$.

REMARK 10.15. We know, by Theorem 8.30 that all the tori in the resonant region are given by formulas (10.7) for $E = E_i, F_i, G$ and verify that

$$
\begin{aligned}
|E_i - E_{i+1}| &\leq \varepsilon^{\frac{3}{2}+1} \leq \max(\sqrt{E_i}, \varepsilon), \\
|E_{l_E} - F_1| &\leq \varepsilon^{\frac{3}{2}+1} \leq \max(\sqrt{E_{l_E}}, \varepsilon), \\
|F_i - F_{i+1}| &\leq \varepsilon^{\frac{3}{2}+1} \leq \max(\sqrt{F_i}, \varepsilon) \\
|F_{l_F} - G| &\leq \varepsilon^{\frac{3}{2}+1} \leq \max(\sqrt{F_{l_F}}, \varepsilon).
\end{aligned}
$$

and that by and Corollary 8.31 and Remark 8.32 that, when they are written as graphs, the distance between to consecutive tori is of order $O(\varepsilon^{3/2})$.

Then, when we apply Lemma 10.14 to these tori we obtain that the image under the scattering map of a torus in this region intersect transversally another torus.

PROOF. It suffices to apply Lemma 10.7 to the function

$$
F(I, \varphi, s) = \frac{y^2}{2}(1 + O(\varepsilon)) + O_{\mathcal{C}^2}(\varepsilon^2),
$$

and $\lambda_E^{\pm}$ given by (10.11) and (10.12).

In order to check inequality (10.3) we use formula (9.9), so that $S = \mathrm{Id} + \varepsilon S_1 + O_{\mathcal{C}^1}(\varepsilon^{1+\varrho})$, and that, by (8.57), $\lambda_E^{\pm}$ is a bounded function and its derivatives with respect to $\varphi, s$ are of order $O(\varepsilon)$.

We will also use that

$$\nabla_{I,\varphi,s} F(I,\varphi,s;\varepsilon) = y(I,\varphi,s;\varepsilon)(1 + O_{\mathcal{C}^2}(\varepsilon))(k_0,0,0) + O_{\mathcal{C}^2}(\varepsilon^2),$$

and analogous estimates for the second derivatives of $F$. Hence, we can compute, as in Lemma 10.11:

$$
\begin{aligned}
F \circ S &\circ (\lambda_E^\pm, \mathrm{Id}, \mathrm{Id})(\varphi, s) \\
&= E - \varepsilon(k_0 + O_{\mathcal{C}^2}(\varepsilon)) f_E^\pm(x,s;\varepsilon) \cdot \\
&\quad \left( \frac{\partial \mathcal{L}^*}{\partial \theta}(\lambda_E^\pm(\varphi,s;\varepsilon), \varphi - \lambda_E^\pm(\varphi,s;\varepsilon)s) + O_{\mathcal{C}^1}(\varepsilon^\varrho) \right) \\
&\quad + k_0^2 \varepsilon^2 \left( \frac{\partial \mathcal{L}^*}{\partial \theta}(\lambda_E^\pm(\varphi,s;\varepsilon), \varphi - \lambda_E^\pm(\varphi,s;\varepsilon)s) + O_{\mathcal{C}^1}(\varepsilon^\varrho) \right)^2 + O_{\mathcal{C}^1}(\varepsilon^3) \\
&= E - \varepsilon k_0 \mathcal{Y}_\pm(\theta, E) \cdot \\
&\quad \left( \frac{\partial \mathcal{L}^*}{\partial \theta}\left(-\frac{l_0}{k_0} + \mathcal{Y}_\pm(\theta, E), \varphi - \left(-\frac{l_0}{k_0} + \mathcal{Y}_\pm(\theta, E)\right)s\right) + O_{\mathcal{C}^1}(\varepsilon^\varrho) \right) \\
&\quad + k_0 \varepsilon^2 \left( \frac{\partial \mathcal{L}^*}{\partial \theta}\left(-\frac{l_0}{k_0} + \mathcal{Y}_\pm(\theta, E), \varphi - \left(-\frac{l_0}{k_0} + \mathcal{Y}_\pm(\theta, E)\right)s\right) \right)^2 \\
&\quad + O_{\mathcal{C}^1}(\varepsilon^{5/2}, \varepsilon^{2+\varrho}) \\
&= E - \varepsilon \mathcal{M}_\pm(\theta;\varepsilon) + O_{\mathcal{C}^1}(\varepsilon^{5/2}, \varepsilon^{2+\varrho})
\end{aligned}
$$

Now, we compute the main part of the functions $\mathcal{M}_\pm$

$$
\begin{aligned}
\mathcal{M}_\pm(\theta;\varepsilon) &= k_0 \mathcal{Y}_\pm(\theta, E) \cdot \\
&\quad \left( \frac{\partial \mathcal{L}^*}{\partial \theta}\left(-\frac{l_0}{k_0} + \frac{1}{k_0}\mathcal{Y}_\pm(\theta, E), \varphi - \left(-\frac{l_0}{k_0} + \frac{1}{k_0}\mathcal{Y}_\pm(\theta, E)\right)s\right) \right. \\
&\quad \left. + O_{\mathcal{C}^1}(\varepsilon^\varrho) \right) \\
&\quad - \varepsilon k_0^2 \left( \frac{\partial \mathcal{L}^*}{\partial \theta}\left(-\frac{l_0}{k_0} + \frac{1}{k_0}\mathcal{Y}_\pm(\theta, E), \varphi - \left(-\frac{l_0}{k_0} + \frac{1}{k_0}\mathcal{Y}_\pm(\theta, E)\right)s\right) \right)^2
\end{aligned}
$$

The analysis of the functions $\mathcal{M}_\pm$ will be done differently whether we are close to the resonance or whether we are reasonably far. We choose any $\nu$ such that $1 \le \nu < 2$, and we will consider the following two cases:

**Close to the resonance:** $-c_4\varepsilon^2 \le E \le \tilde{c}\varepsilon^\nu$.
In this region, we have that $|\ell(\cdot, E)|_{\mathcal{C}^1} \le \varepsilon^{\nu/2}$, and

$$\mathcal{Y}_\pm(\theta, E) = \pm(1 + \varepsilon b)\ell(\theta, E) + \varepsilon \tilde{\mathcal{Y}}_\pm \circ \ell(\theta, E).$$

By Lemma 8.34 we know that $\tilde{\mathcal{Y}}_\pm$ have $\mathcal{C}^2$ norm bounded independently of $\varepsilon$, $E$ and $\tilde{\mathcal{Y}}(0) = \tilde{\mathcal{Y}}'(0) = 0$, so that

$$\left| \varepsilon \tilde{\mathcal{Y}} \circ \ell \right|_{\mathcal{C}^1} \le \mathrm{cte.} \, \varepsilon^{1+\nu}.$$

Then, the main terms in $\mathcal{M}_\pm$ are

$$\mathcal{M}_\pm(\theta;\varepsilon) =$$

$$= \pm k_0\sqrt{2(E-\varepsilon^2 U(\theta;0))}\frac{\partial\mathcal{L}^*}{\partial\theta}(-\frac{l_0}{k_0},\frac{\theta}{k_0}) - \varepsilon k_0^2\left(\frac{\partial\mathcal{L}^*}{\partial\theta}(-\frac{l_0}{k_0},\frac{\theta}{k_0})\right)^2$$

$$+ |\ell|\, O_{\mathcal{C}^1}(\varepsilon^{\nu/2},\varepsilon^\varrho) + O_{\mathcal{C}^1}(\varepsilon^{1+\nu}).$$

To apply Lemma 10.7 we need to check (10.3), that is, it suffices to show that we can bound from below the derivative with respect to $\theta = k_0\varphi + l_0 s$ of this function divided by $k_0\lambda_E^\pm(\varphi,s;\varepsilon) + l_0$. Computing this derivative, we obtain:

$$\frac{\partial}{\partial\theta}(\mathcal{M}_\pm(\varphi,s;\varepsilon))$$

$$= \frac{\pm k_0}{\sqrt{2(E-\varepsilon^2 U(\theta;0))}}\left(2E\frac{\partial^2}{\partial\theta^2}\mathcal{L}^*(-\frac{l_0}{k_0},\frac{\theta}{k_0})\right.$$

$$-\varepsilon^2\left[k_0 U'(\theta;0)\frac{\partial\mathcal{L}^*}{\partial\theta}(-\frac{l_0}{k_0},\frac{\theta}{k_0}) + 2U(\theta;0)\frac{\partial^2}{\partial\theta^2}\mathcal{L}^*(-\frac{l_0}{k_0},\frac{\theta}{k_0})\right]\Big)$$

$$- 2\varepsilon k_0\frac{\partial\mathcal{L}^*}{\partial\theta}(-\frac{l_0}{k_0},\frac{\theta}{k_0})\frac{\partial^2}{\partial\theta^2}\mathcal{L}^*(-\frac{l_0}{k_0},\frac{\theta}{k_0})$$

$$+ |\ell|\, O_{\mathcal{C}^0}(\varepsilon^{\nu/2},\varepsilon^\varrho) + O_{\mathcal{C}^0}(\varepsilon^{1+\nu}).$$

Then,

$$\frac{\frac{\partial}{\partial\theta}(\mathcal{M}_\pm(\theta;\varepsilon))}{k_0\lambda_E^\pm(\varphi,s;\varepsilon) + l_0}$$

$$= \frac{k_0}{2(E-\varepsilon^2 U(\theta;0))}\left(2E\frac{\partial^2}{\partial\theta^2}\mathcal{L}^*(-\frac{l_0}{k_0},\frac{\theta}{k_0})\right.$$

$$-\varepsilon\left[k_0 U'(\theta;0)\frac{\partial\mathcal{L}^*}{\partial\theta}(-\frac{l_0}{k_0},\frac{\theta}{k_0}) + 2U(\theta;0)\frac{\partial^2}{\partial\theta^2}\mathcal{L}^*(-\frac{l_0}{k_0},\frac{\theta}{k_0})\right]$$

$$\mp 2\varepsilon k_0\sqrt{2(E-\varepsilon^2 U(\theta;0))}\frac{\partial\mathcal{L}^*}{\partial\theta}(-\frac{l_0}{k_0},\frac{\theta}{k_0})\frac{\partial^2}{\partial\theta^2}\mathcal{L}^*(-\frac{l_0}{k_0},\frac{\theta}{k_0})\Big)$$

$$+ O_{\mathcal{C}^0}(\varepsilon^{\nu/2},\varepsilon^\varrho).$$

By hypothesis (10.14) we know that the main term of this expression is bounded away from zero by a constant $C$ for $\theta \in \mathcal{J} \subset \mathcal{J}^*_{-l_0/k_0}$. Consequently, the angle of intersection can be bounded again from below by $C'\varepsilon$, for some suitable constant independent of $\varepsilon$.

In order to see which surfaces will intersect $S(L_E^{F,\pm})$, we observe that the function $\mathcal{M}_\pm(\theta;\varepsilon)$ is given approximately by

$$k_0\frac{\partial\mathcal{L}^*}{\partial\theta}(-\frac{l_0}{k_0},\frac{\theta}{k_0})\left(\pm\sqrt{2(E-\varepsilon^2 U(\theta;0))} - \varepsilon k_0\frac{\partial\mathcal{L}^*}{\partial\theta}(-\frac{l_0}{k_0},\frac{\theta}{k_0})\right)$$

and there are different behaviors depending on the branch $S(L_E^{F,\pm})$.

If we focus in the case of $S(L_E^{F,-})$, which corresponds to the lower branch, the function $\mathcal{M}_-$ can have different signs depending on the size of $\sqrt{2(E - \varepsilon^2 U(\theta; 0))}$.

If

$$\sqrt{2(E - \varepsilon^2 U(\theta; 0))} > -\varepsilon k_0 \frac{\partial \mathcal{L}^*}{\partial \theta} \left( -\frac{l_0}{k_0}, \frac{\theta}{k_0} \right)$$

then $\mathcal{M}_-(\theta; \varepsilon) > 0$ and $S(L_E^{F,-})$ will intersect the surfaces $L_{E'}^{F,-}$, with $E' < E$. This happens for positive values of $E$, which correspond to the primary tori under the separatrix loop, and for negative values of $E$, which correspond to the secondary tori inside the separatrix loop.

If $\sqrt{2(E - \varepsilon^2 U(\theta; 0))}$ is smaller than $-\varepsilon k_0 \frac{\partial \mathcal{L}^*}{\partial \theta}(-\frac{l_0}{k_0}, \frac{\theta}{k_0})$ then $\mathcal{M}_-(\theta; \varepsilon) < 0$, we obtain that $S(L_E^{F,-})$ will intersect the surfaces with $E' > E$. This means that, due to the fact that the Melnikov function is big, this surface intersect the surfaces $L_{E'}^{F,+}$. Then, in the case of a resonance of second order it is possible that only with one application of $S$ we cross the gap.

Once we have cross the separatrix loop, that is, when we consider $S(L_E^{F,+})$, we have that $\mathcal{M}_+(\theta; \varepsilon)$ is always negative, so we cross the surfaces $L_{E'}^{F,+}$, for $E' > E$.

In all these cases we have that there exists some constant $C''$ such that

$$\max_{\theta \in \mathcal{J}_1 \cup \mathcal{J}_2 \cup \mathcal{J}_3} |\mathcal{M}(\lambda_E(\varphi, s; \varepsilon), \varphi - \lambda_E(\varphi, s; \varepsilon)s)| \geq C'' \max(E^{1/2}, \varepsilon),$$

and applying Lemma 10.3 we obtain the desired result.

**Far from the resonance:** $\tilde{c}\varepsilon^\nu \leq E \leq c_2 L$.

This case is analogous to the non resonant region, because in this case,

$$\begin{aligned} |\ell(\theta, s)| &= \sqrt{2(E - \varepsilon^2 U(x; \varepsilon))} = \sqrt{2E} \sqrt{1 - \frac{\varepsilon^2}{E} U(x; \varepsilon)} \\ &= \sqrt{2E}(1 + O_{\mathcal{C}^2}(\varepsilon^{2-\nu})), \end{aligned}$$

consequently, as $\sqrt{2E} \geq \varepsilon^{\nu/2} \gg \varepsilon$, the functions $\mathcal{M}_\pm$ become

$$\mathcal{M}_\pm(\theta; \varepsilon)$$
$$= k_0 \sqrt{2E} \frac{\partial \mathcal{L}^*}{\partial \theta} \left( -\frac{l_0}{k_0} + \sqrt{2E}, \varphi - (-\frac{l_0}{k_0} + \sqrt{2E})s \right) + O_{\mathcal{C}^2}(\varepsilon^{2-\nu}).$$

Then, if the function $\frac{\partial \mathcal{L}^*}{\partial \theta}(I, \theta)$ is not constant and negative as a function of $\theta$, we can apply Lemma 10.7, to get the desired result.

$\square$

Hypothesis **H5"'** in (10.14) amounts to the existence of a lower bound for a function of the following type:

$$\frac{a(\theta)E + b(\theta)\varepsilon^2 + c(\theta)\varepsilon\sqrt{E - \varepsilon^2 U(\theta;0)}}{E - \varepsilon^2 U(\theta;0)}$$

Concretely we have

$$a(\theta) = k_0 \frac{\partial^2 \mathcal{L}^*}{\partial \theta^2}\left(-\frac{l_0}{k_0}, \frac{\theta}{k_0}\right)$$

$$(10.15) \quad b(\theta) = -\frac{k_0}{2}\left[k_0 U'(\theta;0)\frac{\partial \mathcal{L}^*}{\partial \theta}\left(-\frac{l_0}{k_0}, \frac{\theta}{k_0}\right) + 2U(\theta;0)\frac{\partial^2 \mathcal{L}^*}{\partial \theta^2}\left(-\frac{l_0}{k_0}, \frac{\theta}{k_0}\right)\right]$$

$$c(\theta) = \frac{\pm k_0}{\sqrt{2}}\frac{\partial^2 \mathcal{L}^*}{\partial \theta^2}\left(-\frac{l_0}{k_0}, \frac{\theta}{k_0}\right)$$

The following Lemma gives a computable sufficient condition that guaranties that hypothesis **H5"'** is verified.

LEMMA 10.16. *Let $a(\theta)$, $b(\theta)$, $c(\theta)$ be functions of class $\mathcal{C}^r$, $r \geq 0$, with $a(\theta) \neq 0$, such that there exist $\theta_1$, $\theta_2$, $\theta_3$ in some interval $\mathcal{J}$ verifying*

$$(10.16) \quad \begin{vmatrix} \tilde{a}(\theta_1) & \tilde{a}(\theta_2) & \tilde{a}(\theta_3) \\ \tilde{b}(\theta_1) & \tilde{b}(\theta_2) & \tilde{b}(\theta_3) \\ \tilde{c}(\theta_1) & \tilde{c}(\theta_2) & \tilde{c}(\theta_3) \end{vmatrix} \neq 0,$$

*where*

$$(10.17) \quad \begin{aligned} \tilde{a}(\theta) &= a(\theta)^2 \\ \tilde{b}(\theta) &= 2a(\theta)b(\theta) - c(\theta)^2 \\ \tilde{c}(\theta) &= b(\theta)^2 - c(\theta)^2 U(\theta;0) \end{aligned}$$

*Then, there exists a constant $\tilde{C}$ and three intervals $\theta_i \in \mathcal{J}_i \subset \mathcal{J}$, $i = 1, 2, 3$, such that given any $(x, y) \in \mathbb{R}^2$, one can choose $i = 1, 2, 3$ and, for $\theta \in \mathcal{J}_i$*

$$(10.18) \quad \left| a(\theta)x + b(\theta)y^2 + c(\theta)y\sqrt{x - y^2 U(\theta;0)} \right| \geq \tilde{C}\sqrt{x^2 + y^4}.$$

PROOF. We call $z = \frac{x}{y^2}$, and consider the function

$$f(\theta, z) = \frac{a(\theta)z + b(\theta) + c(\theta)\sqrt{z - U(\theta;0)}}{\sqrt{1 + z^2}} = \frac{g(\theta, z)}{\sqrt{1 + z^2}}$$

Then, it is enough to prove that there exists a constant $C$ and three intervals $\theta_i \in \mathcal{J}_i$, $i = 1, 2, 3$, such that given any value of $z \in [c_1, \infty)$, one can choose $i = 1, 2, 3$ and, for $\theta \in \mathcal{J}_i$

$$(10.19) \quad |f(z)| \geq \tilde{C}$$

One can check easily that condition (10.16) implies that there is no $z$ such that the functions $g(\theta_i, z)$ vanish simultaneously. Then, given any value of $z$ and calling

$$c(z) = \max(|f(\theta_1, z)|, |f(\theta_2, z)|, |f(\theta_3, z)|)$$

we have that $c(z) > 0$. Moreover as $\lim_{z \to \infty} f(\theta_i, z) = a(\theta_i) \neq 0$, we have that $\min c(z) \geq \tilde{C} > 0$, then, for any value of $z \in [c_1, \infty)$, $c(z) \geq C$, and this concludes the proof.

$\square$

## 10.3. Existence of transition chains to objects of different topological types

In this section, we prove for $0 < \varepsilon \ll 1$ the existence of heteroclinic trajectories between different kinds of invariant manifolds (primary tori, secondary tori, or periodic orbits). These transitions chains have a span of order 1 in $I$ and may go over the resonant regions which are devoid of primary KAM tori.

The precise formulation of the result is given in Proposition 10.17.

Notice that, by the assumption that we are dealing with a polynomial perturbation, we have only a finite number of resonances of order $1, 2$ in the interval $(I_-, I_+)$.

PROPOSITION 10.17. *Consider a Hamiltonian (3.3) satisfying the hypotheses of Theorem 4.1.*

*Let $\{\hat{I}_i\}_{i=1}^M \subset (I_-, I_+)$ be the set of resonaces of order 1 or 2.*

*For each of these resonances select $\hat{T}_i$ be either a secondary torus among those produced in Theorem 8.30, or the weak (un)stable manifolds of a periodic orbit as constructed in Proposition 8.40.*

*Pick two KAM tori $\mathcal{T}_\pm$ away from the resonances of order $1, 2$ and such that $|I(\mathcal{T}_\pm) - I_\pm| \leq \varepsilon^{3/2}$. (These tori exist because of Theorem 8.12).*

*Then, there exists a transition chain $\{\mathcal{T}_i\}_{i=0}^{N(\varepsilon)}$ (we can take $N(\varepsilon) \leq C/\varepsilon$) in such a way that*

a) *The transition chain is obtained through applications of the scattering map. That is.*

$$S(\mathcal{T}_i) \pitchfork \mathcal{T}_{i+1}$$

b) $\mathcal{T}_0 = \mathcal{T}_-$, $\mathcal{T}_{N(\varepsilon)} = \mathcal{T}_+$.

c) *The transition chain contains all the $\hat{T}_i$ selected above.*

REMARK 10.18. The reason to include the conclusion a) above is that this can affect some estimates of the speed of diffusion.

As it is well known—and we will see in the next chapter—any subsequence of a transition chain is a transition chain, so that, in general, it is not easy to speak about the number of tori in a transition chain starting somewhere and ending somewhere else. In particular, there is no hope of estimating the time needed to transverse a transition chain by the number of elements unlesss one restricts somehow the mechanisms by which the transistion is generated. The conclusion a) gives a natural way of identifying chains all whose steps are obtained through a similar mechanism.

PROOF. The proof is based on collecting the results developed in Section 10.2.

We note that, because of the Lemma 10.8, $S(\mathcal{T}_0)$ intersects all the primary KAM tori for which the averaged energy $E$ lies in an interval $\mathcal{I}_1 \equiv (E_0 + C\varepsilon, E_0 + C''\varepsilon)$. Therefore, by Lemma 10.4 the torus $\mathcal{T}_0$ has a heteroclinic connection with all the tori of energy in the interval $\mathcal{I}_1$. We also recall that it was shown in Proposition 8.21 and Theorem 8.30 that in $\mathcal{I}_1$ we can find tori with gaps at most $\varepsilon^{3/2}$. Hence, we obtain that the torus $\mathcal{T}_0$ has heteroclinic intersections with KAM primary tori that cover the interval $\mathcal{I}_1$ with gaps $\varepsilon^{3/2}$.

Starting with a torus $\mathcal{T}_{E'}$ of energy $E' \in \mathcal{I}_1$ we see that $S(\mathcal{T}_{E'})$ intersects transversally all the primary KAM tori whose energy lies in an interval $(E' + C\varepsilon, E' + C''\varepsilon)$. Hence, we see that the torus $\mathcal{T}_0$ has heteroclinic connections with all the tori whose energy lies in an interval $\mathcal{I}_2 = (E_0 + 2C\varepsilon, E_0 + 2C''\varepsilon)$. Repeating the argument $K$ times, as long as we are in the non-resonant region $[I_0, \hat{I}_1 - L]$, we see that the torus $\mathcal{T}_0$ has heteroclinic connections with all the KAM primary tori whose energy lies in the interval $\mathcal{I}_K = (E_0 + KC\varepsilon, E_0 + KC''\varepsilon)$. We note if we take $K \geq K_* := C/(C'' - C)$ we see that $\mathcal{I}_K \cap \mathcal{I}_{K+1} \neq \emptyset$. Therefore, choosing $K$ big enough, we conclude that the torus $\mathcal{T}_0$ has heteroclinic connections with all the KAM primary tori in the region $(I_0 + K_* C\varepsilon, \hat{I}_1 - L)$.

Let $\hat{\mathcal{T}}_1$ be the a secondary torus among those produced in Theorem 8.30, (or the weak (un)stable manifold of a periodic orbit as constructed in Proposition 8.40) selected in the statement of the current proposition.

Note that $\hat{\mathcal{T}}_1$ is included in the resonant region $[\hat{I}_1 - 2L, \hat{I}_1 + 2L] \times \mathbb{T}^2$ (which overlaps with the non-resonant region $[\hat{I}_0, \hat{I}_1 - L] \times \mathbb{T}^2$). We have shown in Lemma 10.11, for a resonance of order one, and in Lemma 10.14, for a resonance of order two (and more precilsely in remarks 10.12 and 10.15), that we can find a transition chain which connects $\hat{\mathcal{T}}_1$ with some KAM primary torus $\mathcal{T}_1^*$ whose equation is $I = I^* + \mathrm{O}(\varepsilon)$, with $I^* \in [\hat{I}_1 - 2L, \hat{I}_1 - L] \subset [I_0 + K_* C\varepsilon, \hat{I}_1 - L]$, hence, we can construct a piece of a chain that starts in $\mathcal{T}_0$ and reaches all the way to $\mathcal{T}_1^*$.

We have also shown in Lemma 10.11 and Lemma 10.14 and in remarks 10.12 and 10.15 that we can find a transition chain that connects $\hat{\mathcal{T}}_1$ and $\tilde{\mathcal{T}}_1$, where $\tilde{\mathcal{T}}_1$ is a KAM primary torus whose equation is $I = \hat{I}_1 + L + \mathrm{O}(\varepsilon)$. (Recall that for a resonance of order two, an alternative transition chain may be constructed for a primary torus $\hat{\mathcal{T}}_1$ in the non resonant region $(\mathcal{I}_0, \hat{I}_1)$, as described in Lemma 10.14.)

Again we can argue that the torus $\tilde{\mathcal{T}}_1$ is connected to all the tori in the region $(\hat{I}_1 + L + K_* C\varepsilon, \hat{I}_2 - L)$, and we can again repeat the process to see that we traverse all the transition chain. $\qquad \square$

We note that if we have defined different scattering maps—or the same scattering map in a different domain—we can construct the transition chains using any of the scattering maps.

In particular, if we have at the same time scattering maps defined in regions which increment the $I$ and in regions where the $I$ is decreased, we can construct infinitely transition chains that make largely arbitrary excursions in the variable $I$.

CHAPTER 11

# Orbits shadowing the transition chains and proof of theorem 4.1

In this chapter, we just state the well known result that given a transition chain, we can find an orbit visiting all the elements of the chain.

There are many proofs of similar results in the literature. In our case, however, we have to make sure that the proofs remain valid for transition chains that incorporate objects with different topologies and indeed different dimensions. In our formulation also we allow the transition chains to have infinite many transition tori. This makes certain well known proofs in the literature not applicable. Hence, we present full details, following the exposition in [**DLS00**]. An exposition of related results along similar lines can be found in [**FM03**].

LEMMA 11.1. *Let* $\{\mathcal{T}_i\}_{i=0}^N$ *be a transition chain.*

*Given* $(\delta_i)_{i=0,\dots,N}$ *be a sequence of strictly positive numbers, we can find a point* $\tilde{x} \in (\mathbb{R} \times \mathbb{T})^2 \times \mathbb{T}$, *and a increasing sequence of numbers* $0 = t_0 < \dots < t_N$ *such that*

$$\tilde{\Phi}_{t_i,\varepsilon}(\tilde{x}) \in \mathbf{B}(\mathcal{T}_i, \delta_i)$$

*where* $\mathbf{B}(\mathcal{T}_i, \delta_i)$ *is neighborhood of size* $\delta_i$ *of the torus* $\mathcal{T}_i$.

A particular case of the results of [**FM00**] is:

LEMMA 11.2. *Let* $f$ *be a* $\mathcal{C}^2$ *symplectic mapping in a symplectic manifold. Assume that the map leaves invariant* $\mathcal{C}^1$ *torus* $\mathcal{T}$ *and that the motion on the torus is an irrational rotation. Let* $\Gamma$ *be a manifold intersecting* $W_{\mathcal{T}}^{\mathrm{u}}$ *transversally. Then,*

$$W_{\mathcal{T}}^{\mathrm{s}} \subset \overline{\bigcup_{i>0} f^{-i}(\Gamma)}$$

We emphasize that since the proof in [**FM00**] only makes assumptions about the motion on the torus and the linearization around it, it applies independently of whether the torus is primary or secondary. Also, it is independent of the dimension.

In the case that the transition chain is finite, this establishes the desired result simply by the continuous dependence on the initial conditions.

If we were interested in getting infinitely long trajectories that perform abitrary excursions, we would need to deal with infinitely long transition chains. Not all the arguments in the literature can deal with infinitely long transition chains and some papers emphasize that. Nevertheless, there is a very simple minded point set topology argument that allows to deal with infinitely long transition chains. We reproduce the argument verbatim from [**DLS00**].

LEMMA 11.3. *Let $\{\mathcal{T}_i\}_{i=1}^{\infty}$ be a sequence of transition tori. Given $\{\varepsilon_i\}_{i=1}^{\infty}$ a sequence of strictly positive numbers, we can find a point $P$ and a increasing sequence of numbers $T_i$ such that*

$$\Phi_{T_i}(P) \in N_{\varepsilon_i}(\mathcal{T}_i)$$

*where $N_{\varepsilon_i}(\mathcal{T}_i)$ is a neighborhood of size $\varepsilon_i$ of the torus $\mathcal{T}_i$.*

PROOF. Let $x \in W^{\mathrm{s}}_{\mathcal{T}_1}$. We can find a closed ball $B_1$, centered on $x$, and such that

(11.1)                    $\Phi_{T_1}(B_1) \subset N_{\varepsilon_1}(\mathcal{T}_1).$

By the Inclination Lemma 11.2

$$W^{\mathrm{s}}_{\mathcal{T}_2} \cap B_1 \neq \emptyset.$$

Hence, we can find a closed ball $B_2 \subset B_1$, centered in a point in $W^{\mathrm{s}}_{\mathcal{T}_2}$ such that, besides satisfying (11.1):

$$\Phi_{T_2}(B_2) \subset N_{\varepsilon_2}(\mathcal{T}_2).$$

Proceeding by induction, we can find a sequence of closed balls

$$B_i \subset B_{i-1} \subset \cdots \subset B_1$$
$$\Phi_{T_j}(B_i) \subset N_{\varepsilon_j}(\mathcal{T}_j), \quad i \leq j.$$

Since the balls are compact, $\cap B_i \neq \emptyset$. A point $P$ in the intersection satisfies the required property. $\qquad\square$

END OF THE PROOF OF THEOREM 4.1. In order to prove the existence of an orbit $\tilde{x}(t) = \tilde{\Phi}_{t,\varepsilon}(\tilde{x})$ of system (3.3) verifying (4.7) we only need to apply Lemma 11.1 to the transition chains obtained in proposition 10.17.

Then, Theorem 4.1 is proved.

$\qquad\square$

CHAPTER 12

# Conclusions and remarks

## 12.1. The role of secondary tori and the speed of diffusion

We have shown that the secondary whiskered tori as well as lower dimensional tori, i.e. periodic orbits, can be used in the construction of transition chains to overcome the *"large gap problem"* in the study of diffusion.

We think that this is a step toward reconciling the physical literature and intuition [**Chi79, Ten82**] (that states that the main cause of instability and diffusion are resonances) and the mathematical literature, which hitherto emphasized methods that were based in the mechanism of [**Arn64**] which requires that KAM tori are close and, hence, that there are no resonances.

We think that it will be quite interesting to understand better the geometric objects that one can find in different types of resonances.

## 12.2. Comparison with [DLS00]

Even if this paper and [**DLS00**] are logically independent, there are several analogies in the general strategy and the tools of the present work and that in [**DLS00**].

Nevertheless, there are several important differences in the geometry of the systems considered here and that in [**DLS00**] (and in [**Mat95, BT99**]).

Let us emphasize the most salient geometric differences for the reader interested in comparing the two papers:

- The manifolds $W^{\mathrm{s}}_{\tilde{\Lambda}}$ and $W^{\mathrm{u}}_{\tilde{\Lambda}}$ considered in this paper do not intersect transversely in the unperturbed case. This is why we refer to the system here as a priori unstable, but we coined the name *a priori chaotic* for the system in [**Mat95, DLS00, BT99, DLS05**]. This has made it necessary for us to study the transverse intersection through a perturbation theory.

- The homoclinic connections of the model here do not include a phase shift. (See equation (6.7) and compare it with equation (2.1) of [**DLS00**].)

  In the case of [**Mat95, DLS00, BT99**], there was a phase shift and, hence, we needed to compare the KAM tori at different points. This required a more detailed KAM perturbation theory carried out in Lemma 4.16 of [**DLS00**]. Such considerations will not be needed in this paper, but could be reintroduced to analyze a more complicated model.

123

- The most important difference with [**DLS00**] is that in our problem there is no analogue of the scaling (Section 4.2 in [**DLS00**]) which makes the system, for high energy, to become fast with respect to the perturbation. Hence, the fact that gaps between tori in [**DLS00**] were very small has not been maintained in our case and we had deal with the so called *large gap problem*.

Of course, the fact that models with very different features can be understood with similar tools makes us quite hopeful that even more complicated models can be analyzed in similar ways.

## 12.3. Heuristics on the genericity properties of the hypothesis and the phenomena

The verification we have presented of the mechanism is geared towards the explicit verification of the mechanism in concrete systems.

This point of view is motivated because in applications or in mathematics, one has to deal with concrete systems—e.g. the solar system or the systems appearing in technology—which have very special properties.

Nevertheless, several colleagues have asked about the genericity of the mechanism that we have discussed here.

In this section, we will discuss briefly and informally the genericity hypothesis of Theorem 4.1.

First note that the hypotheses **H1** and **H2** on the pendulum hold for an open and dense set of penduli (indeed for all but a manifold of infinite codimension) in any $\mathcal{C}^r$ topology, $r \geq 1$.

Hence, we will consider the genericity of the hypothesis on $h$ in the space of trigonometric polynomials—this is Hypothesis **H3**—once we fix the pendulum satisfying **H1**, **H2**. We note that the pendulum will have two connecting orbits that are geometrically different.

If we fix one of the connecting orbits, we note that the mapping that sends the perturbation to the Melnikov potential (4.3) is linear and nontrivial. In this case, generic properties on the Melnikov potential are implied by generic assumptions on the perturbation. Moreover, since $h$ is assumed to be a polynomial, we do not need to discuss in which $\mathcal{C}^r$ genericity properties hold.

We note that it is generic that there are open and non-empty sets for which we have **H4** which include the resonances in $\tilde{\Lambda}_\varepsilon$.

This follows because the existence of such an open set is clearly open in the space of $h$'s. The density follows because the negation of the conclusion is that for all points in $\tilde{\Lambda}_\varepsilon$ in a resonance $\{k/l\} \times \mathbb{T}^2$, the function $\Gamma$ has only degenerate critical points. This can be easily destroyed by arbitrarily small perturbations.

Hence, we conclude that the set of models of the form (3.3) that overcome the large problem *in one resonance* is open and dense. We emphasize that in this argument dealing with one resonance Hypothesis **H3** does not play any role.

We can also wonder about the genericity of the existence of orbits that transverse all the resonances. The following discussion, based simply on the counting of parameters, presents some plausibility arguments for what one would expect.

Notice that given a critical point for $\Gamma$ we expect that it will be non-degenerate except for a set of $I, \varphi, s$ of codimension 1. In these sets, we will have that it is generically possible to find sets $H_-^{\pm}$ where we have different signs in (4.6). Generically, these sets will have overlapping projections on the $I$ direction, hence, it will be possible to obtain the existence of a symbolic dynamics.

If we consider two critical points, we expect that the sets of $I, \varphi, s$ for which one of them are non-degenerate is a set of codimension 2.

Furthermore, if we consider systems for which the pendulum has two homoclinics, we have more critical points to study. Once we have 4 or more we expect that all the points in $\tilde{\Lambda}_\varepsilon$ correspond to a non-degenerate critical point.

Hence, we expect that the intervals corresponding to a positive sign in (4.6) overlap and their union covers all the resonant regions. Similarly, the intervals the negative signs in (4.6) will overlap and cover the resonances. Hence, we expect that, for the systems we have consider it will be generic to have trajectories that cross all the resonances.

We postpone a detailed verification of the results for generic systems. We hope that a generic verification of the system could be simpler and applicable without **H3**.

Note that all our results on existence of orbits changing the action are formulated for all $\varepsilon$ with $|\varepsilon| \leq \varepsilon^*$ with $\varepsilon^* > 0$. The $\varepsilon^*$, however, depends on $h$ and is only defined for $h$ that satisfy some non-degeneracy conditions. As $h$ approaches a degenerate $h^*$, $\varepsilon^*(h)$ approaches zero.

If we do not want to consider parameters, we are lead to consider the set of $\varepsilon h$ for which we can establish diffusion across the big gaps. This set contains intervals in $\varepsilon$, but the size of the interval can go to zero as we approach the degenerate sets. This seems to be very similar to the *"cusp residual"* sets considered in [**Mat02**].

## 12.4. The hypothesis of polynomial perturbations

We observe that in the proof we have never used really that the perturbation is a polynomial. We have used only that there are a finite number of resonances so that we could use just a uniform distance $L$, independent of $\varepsilon$, among the resonances to isolate the resonances. It seems quite possible that **H3** can be eliminated completely at the price of just carrying out more delicate estimates or claiming only generic results. For the moment, we thought it better to avoid this complication. In the following, we outline a sketch of a heuristic argument for resonances of order 1. For the sake of clarity, we ignore constants. Note the similarity of the argument presented with the heuristic arguments of [**Chi79**] about the resonance overlap criterion.

We note that, because the scattering map moves by an amount of order $\varepsilon$, we only need to consider resonant regions whose size is bigger or equal than $\varepsilon^{1+\delta}$ for an arbitrary positive $\delta$.

A more careful analysis shows that the size of a resonance of order 1 and denominator $k$ can be bounded from above by $(\varepsilon|a_k|)^{1/2}$ where $a_k$ is the coefficient of the Fourier coefficient of the perturbation.

If the perturbation is $C^\ell$, we have that $|a_k| \leq k^{-\ell}$. So that the size of a resonant region can be bounded from above by $\varepsilon^{1/2}|k|^{-\ell/2}$.

Hence, it is clear that it is only necessary to consider resonant regions with denominator $k \leq k_{\max}$ where $k_{\max}$ is such that

$$\varepsilon^{1/2}|k_{\max}|^{-\ell/2} = \varepsilon^{1+\delta}.$$

That is,

$$|k_{\max}| = \varepsilon^{-1/\ell - 2\delta/\ell}$$

We note that the minimum distance between two rational numbers of denominator smaller that $k_{\max}$, and therefore, the minimum distance between the resonances that need to be considered is:

$$|k_{\max}|^{-2} = \varepsilon^{2/\ell + 4\delta/\ell}$$

The maximum width of any of these resonant regions is $\varepsilon^{1/2}$.

Hence, we see that a condition that ensures that the resonant regions are separated is $\ell > 4$.

A similar analysis works for the resonances of order 2.

Once that we have that the resonant regions are separated, we can perform a very similar analysis than the one performed here. The main difference is that rather than choosing $L$ as we have chosen here, one would need to take $L$ depending on $\varepsilon$ and $k$ in such a way that $L$ is much larger than the size of the resonant region. If $\ell$ is large enough, this can be accomplished e.g. by taking $L = \varepsilon^{1/4}|k|^{-\ell/4}$.

The dependence of the averaging results on $L$ can be worked out.

The heuristic argument above is closely related to arguments in [**Chi79**], where it was concluded that for $C^4$ small perturbations, resonances are isolated and one could expect the Twist theorem to hold.

The argument above concludes that when resonances do not overlap, in the sense of [**Chi79**], then the method of this paper applies.

On the other hand, we remark that from the intuition gathered in [**Chi79**], it seems that when the resonances overlap, the diffusion should be more intense.

## 12.5. Involving other objects

We also note that it should be possible to consider other orbits besides the whiskered tori we have used. In particular, given that in a generic system there are hyperbolic periodic orbits approximating the KAM tori (see e.g. [**FL92**]) and that the stable and unstable manifolds are close to the torus, it seems possible that, at least for generic systems one could construct hyperbolic orbits that connect.

It also seems plausible that one could adapt the method of [**Lla02**] using normally hyperbolic laminations to discuss the problem.

More tentatively, it looks plausible that Aubry-Mather Cantor sets and their invariant manifolds could be used as transition elements using arguments similar to those in [**Mat93**].

## 12.6. Variational methods

There is an uncanny relation between the geometric methods and variational methods. We think that it would be very useful to pursue the study of the parallels between the two very different methods. It seems plausible that if one proved connecting lemmas based on variational methods, one could also use Aubry-Mather sets in place of the tori. Of course, variational methods seem to have to use positivity or convexity properties that are not present in our methods.

We note that it is quite possible that one can use a mixed approach. Once one identifies the relevant geometric objects and produce heteroclinic connections among them, variational methods can produce very effective shadowing orbits [**Mat93, CP02, RS02**]. Some implementations of these mixed methods happen in [**Bes96, BCV01**]

## 12.7. Diffusion times

We notice that, during this paper, nothing is said about the diffusion time since the shadowing Lemma 11.1 is only based in topological methods and does not provide quantitative estimates about the ergodization time as other methods do ([**Tre02a, CG03, BB02, BBB03**]). We also note the methods of [**Lla02**] using hyperbolic laminations, that yield very good times. Of course, all these mechanisms have orbits that are very different.

While this paper was under editorial consideration, part of the arguments of the paper were suplemented with the use of the windowing mechanism [**GL05**], which produces explicit estimates on the time for the orbits considered in this paper.

CHAPTER 13

# An example

Consider the Hamiltonian

$$(13.1) \qquad H_\varepsilon(p,q,I,\varphi,t) = \pm \left( \frac{p^2}{2} + \cos q - 1 \right) + \frac{I^2}{2} + \varepsilon \cos q \, g(\varphi,t),$$

where

$$g(\varphi,t) = \sum_{(k,l)\in\mathcal{N}} a_{k,l}\cos(k\varphi+lt) + b_{k,l}\sin(k\varphi+lt)$$

is a trigonometric polynomial in the angles $\varphi$, $t$ ($\mathcal{N}$ is a finite set of indexes). The Hamiltonian of one degree of freedom $P_\pm(p,q) = \pm \left(p^2/2 + \cos q - 1\right)$ is the standard pendulum when we choose the $+$ sign, and its separatrix for positive $p$ is given by (6.2)

$$q_0(t) = 4\arctan e^{\pm t}, \quad p_0(t) = 2/\cosh t.$$

An important feature of the Hamiltonian (13.1) is that the 3-dimensional hyperbolic invariant manifold

$$\tilde{\Lambda} = \{(0,0,I,\varphi,s) : (I,\varphi,s) \in \mathbb{R} \times \mathbb{T}^2\}$$

is *preserved* for $\varepsilon \neq 0$: $p = q = 0 \Rightarrow \dot{p} = \dot{q} = 0$. However, in contrast with the example in [**Arn63b**], the perturbation does not vanish on $\tilde{\Lambda}$. Indeed, restricted to $\tilde{\Lambda}$, the reduced Hamiltonian takes the form $I^2/2 + \varepsilon g(\varphi,t)$. Hence, 2-dimensional whiskered tori

$$\mathcal{T}_I^0 = \{(0,0,I,\varphi,s) : (\varphi,s) \in \mathbb{T}^2\}$$

are not preserved, and primary resonances take place at $I = -l/k$ for each $(k,l) \in \mathcal{N}, k \neq 0$ such that $a_{k,l} \neq 0$. Therefore, (13.1) presents the large gap problem.

The Melnikov potential (4.3) of the Hamiltonian (13.1) is given by

$$\mathcal{L}(I,\varphi,s) = \frac{1}{2} \int_{-\infty}^{\infty} p_0^2(\sigma)g(\varphi+I\sigma,s+\sigma)d\sigma,$$

and computing the integrals by the residue theorem, we obtain:

$$\mathcal{L}(I,\varphi,s) = \sum_{(k,l)\in\mathcal{N}} A_{k,l}(I)\cos(k\varphi+ls) + B_{k,l}(I)\sin(k\varphi+ls),$$

with

$$A_{k,l} = 2\pi \frac{(kI+l)}{\sinh\frac{\pi}{2}(kI+l)} a_{k,l}, \quad B_{k,l} = 2\pi \frac{(kI+l)}{\sinh\frac{\pi}{2}(kI+l)} b_{k,l}.$$

As we will verify in concrete examples, for a very general choice of the coefficients $a_{k,l}, b_{k,l}$ (i.e. for an open dense set of the $a_{k,l}, b_{k,l}$ obtained removing a finite collection of zero sets of analytic—or even algebraic—functions) we can find open sets of $(I, \varphi, s) \in [I_-, I_+] \times \mathbb{T}^2$, such that the function $\tau \in \mathbb{R} \mapsto \mathcal{L}(I, \varphi - I\tau, s - \tau)$ has non-degenerate critical points at a $\tau = \tau^*(I, \varphi, s)$ which verify hypothesis **H4**.

For instance, let us consider the case of a function $g$ with only two harmonics

$$(13.2) \qquad\qquad g(\varphi, t) = a_0 \cos(\varphi) + a_1 \cos(\varphi - t)$$

which gives rise to two "large gaps" associated to the two primary resonances (5.4) $I = 0, 1$, as well as to the "big gap" associated to the secondary resonance (5.5) $I = 1/2$. Assuming that

$$(13.3) \qquad\qquad a_0 a_1 (a_0^2 - a_1^2) \neq 0$$

and choosing, for instance, $[I_-, I_+] = [-1/2, 3/2]$, we are going to check that for all $(I, \varphi, s) \in [I_-, I_+] \times \mathbb{T}^2$, the function $\tau \in \mathbb{R} \mapsto \mathcal{L}(I, \varphi - I\tau, s - \tau)$ has non-degenerate critical points.

First, notice that

$$\mathcal{L}(\tau) := \mathcal{L}(I, \varphi - I\tau, s - \tau) = A_0 \cos(\varphi - I\tau) + A_1 \cos(\varphi - s - (I-1)\tau),$$

with

$$A_0 = A_0(I) = 2\pi \frac{I}{\sinh \frac{\pi}{2} I} a_0, \quad A_1 = A_1(I) = 2\pi \frac{(I-1)}{\sinh \frac{\pi}{2}(I-1)} a_1,$$

so, fixed $(I, \varphi, s)$, we only need to study the evolution of the Melnikov potential

$$(13.4) \qquad\qquad \mathcal{L}(I, \psi_0, \psi_1) = A_0 \cos(\psi_0) + A_1 \cos(\psi_1),$$

expressed in the variables $(\psi_0, \psi_1) = (\varphi, \varphi - s) \in \mathbb{T}^2$, along the straight lines $R = R(I, \varphi, s)$ on the torus:

$$(13.5) \qquad\qquad \psi_0(\tau) = \varphi - I\tau, \quad \psi_1(\tau) = \varphi - s - (I-1)\tau.$$

As long as $a_0 a_1 \neq 0$ (and therefore $A_0 A_1 \neq 0$), the Melnikov potential (13.4) possesses four non-degenerate critical points: a maximum, a minimum and two saddles. Around the two extremum points, its level curves are closed (and indeed convex) curves which fill out a basin ending at the level curve of one of the saddle points.

Therefore, any straight line (13.5) that enters into some extremum basin is tangent to one of the convex closed level curves, giving rise to a non-degenerate extremum of $\mathcal{L}(\tau)$. So, degenerate extrema of $\mathcal{L}(\tau)$ can only exist for straight lines (13.5) that *never* enter inside such extremum basins. In particular, this could only happen for rational values of $I$, since an irrational value of $I$ implies a dense straight line (13.5) (and infinite non-degenerate extrema for $\mathcal{L}(\tau)$).

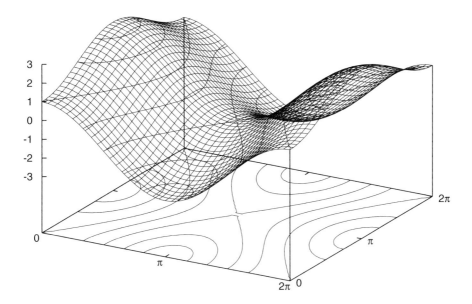

FIGURE 13.1. Graph and level curves of the Melnikov potential $\mathcal{L}(I, \psi_0, \psi_1) = A_0 \cos(\psi_0) + A_1 \cos(\psi_1)$ for $A_0 > -A_1 > 0$. There are four non-degenerate critical points $\psi_0 = 0, \pi$, $\psi_1 = 0, \pi$ and $\mathcal{L}(I, \pi, 0) = -A_0 + A_1 < \mathcal{L}(I, \pi, \pi) = -A_0 - A_1 < \mathcal{L}(I, 0, 0) = A_0 + A_1 < \mathcal{L}(I, 0, \pi) = A_0 - A_1$.

A closer look at the Melnikov potential (13.4) shows that both extremum basins contain a vertical one-dimensional torus (for $|A_0| \geq |A_1|$) or a horizontal one-dimensional torus (for $|A_0| \leq |A_1|$), except for a saddle point. See Figure 13.1 for a pictorial representation of the case $A_0 > -A_1 > 0$. In particular, this means that, except for the case of vertical lines (for $|A_0| \geq |A_1|$), horizontal lines (for $|A_0| \leq |A_1|$), or straight lines joining the two saddle points, every straight line $R$ enters inside both extremum basins. So we already know that, except for the resonant values $I = 0, 1, 1/2$ corresponding to the slopes $\infty, 0, \pm 1$, respectively, of the straight lines (13.5), $\mathcal{L}(\tau)$ has non-degenerate extrema.

On the other hand, for $I = 0, 1, 1/2$ one can check directly that degenerate critical points of $\mathcal{L}(\tau)$ can only appear for $A_0(0) = 0, A_1(1) = 0, A_0(1/2) = \pm A_1(1/2)$, respectively, which is equivalent to $a_0 = 0, a_1 = 0, a_0 = \pm a_1$, respectively.

Therefore, as long as condition (13.3) for the coefficients $a_0, a_1$ is assumed, the function $\tau \in \mathbb{R} \mapsto \mathcal{L}(I, \varphi - I\tau, s - \tau)$ has non-degenerate extrema, and for every $I$ we can find a smooth function $\tau = \tau^*(I, \varphi, s)$ defined in an open set of $(\varphi, s) \in \mathbb{T}^2$.

Moreover, since $\mathcal{L}$ is non-constant along any straight line, it is clear that $\partial_\varphi \mathcal{L}^*$ is non-constant, and besides, for every $I$, there exists a nonempty set $\mathcal{J}_I$ where $\partial_\varphi \mathcal{L}^* < 0$ (and a nonempty set $\mathcal{J}_I^-$ where $\partial_\varphi \mathcal{L}^* > 0$), so hypothesis **H4** is fulfilled.

The conditions **H5'**, **H5"**, **H5'''** can also be checked directly in the example (13.2) at the resonances $I = 0, 1, 1/2$, since we know explicitly the reduced Hamiltonian $I^2/2 + \varepsilon g(\varphi, t)$ given in Proposition 8.4 so that:

$$\varepsilon K_1(J, \varphi, s; \varepsilon) = \varepsilon g(\varphi, s).$$

For example, at $I = 0$ (which comes from the harmonic $k_0 = 1$, $l_0 = 1$), we have, by formula (8.11), $U^{1,0} = a_0 \cos\theta$, $\theta = \varphi$ and Hypothesis **H5'** is verified because $U^{1,0}$ has a non-degenerate critical point at $\theta = 0$. By formulas (8.32), (8.35) and (8.37) we obtain the main term of the Hamiltonian in the resonant region close to $I = 0$:

$$U(x, 0) = a_0(\cos x - 1), \quad x = \theta.$$

On the other hand, for $I = 0$, we have:

$$\mathcal{L}(I, \varphi - I\tau, s - \tau) = A_0(0)\cos\varphi + A_1(0)\cos(\varphi - s + \tau)$$

where $A_0(0) = 4a_0$, $A_1(0) = \frac{2\pi}{\sinh(\pi/2)}a_1$. Therefore $\tau^*(0, \varphi, s) = \varphi - s$ or $\tau^*(0, \varphi, s) = \varphi - s + \pi$, so that the reduced Poincaré function (9.7) reads

$$\mathcal{L}^*(0, \theta) = A_0(0)\cos\theta \pm A_1(0), \quad \theta = \varphi,$$

whose extrema are non-degenerate by condition (13.3).

Then condition **H5"** is also satisfied because

$$a_0 \frac{-\sin^2\theta + 2(\cos\theta - 1)\cos\theta}{2\cos\theta}$$

is non-constant.

Analogous verifications can be carried out in the other resonances.

Hence, we conclude applying Theorem 4.1.

PROPOSITION 13.1. *The Hamiltonian (13.1) with $g$ as in (13.2), where $a_0, a_1$ are such that they are not in the codimension 1 surface of equation $a_0 a_1(a_0^2 - a_1^2) = 0$, and $|\varepsilon| \leq \varepsilon^*(a_0, a_1)$, admits orbits following the mechanism described in this paper and such that $I(0) \leq -1/2$, $I(T) \geq 3/2$ for some $T > 0$.*

We note that Proposition 13.1 is extremely conservative. We have only used the critical points close to the extrema of $\mathcal{L}$. It is clear that there are many other critical points.

The example (13.1) is somewhat non-generic because it has a symmetry that causes that the two homoclinic orbits of the pendulum give the same

Melnikov function. Nevertheless, in spite of the fact that the system is very non-generic, we can verify easily the conditions of our theorem.

# Acknowledgments

We thank J. Vaaler for several discussions and for help with the proof of Lemma 8.17. R.L. thanks B. Fayad for the opportunity to give a series of lectures based on this paper in Paris 13 summer 2004 while this paper was under editorial consideration. We are very grateful to M. Berti, H. Eliasson, B. Fayad, M. Gidea, G. Haller, G. Huguet, V. Kaloshin, J. Mather, P. Roldán and E. Valdinoci, who pointed us typos and suggested improvements and clarifications.

This work has been supported by the *Comisión Conjunta Hispano Norteamericana de Cooperación Científica y Tecnológica*. The final version was prepared while R.L. was enjoying a *Cátedra de la Fundación FBBV*, and A.D. was visiting the *Centre de Recerca Matemàtica*, for whose hospitality he is very grateful. A.D. and T.S. have also been partially supported by the Catalan grant 2001SGR-70, the Spanish grant BFM2000-0805-C02 and the INTAS grant 00-221, and R.L. by NSF grants.

# Bibliography

[AA67]   V. Arnold and A. Avez. *Ergodic problems of classical mechanics.* Benjamin, New York, 1967.

[AKN88]  V. Arnold, V. Kozlov and A. Neishtadt. *Dynamical Systems III*, volume 3 of *Encyclopaedia Math. Sci.* Springer, Berlin, 1988.

[Arn63a] V. I. Arnol'd. Proof of a theorem of A. N. Kolmogorov on the invariance of quasi-periodic motions under small perturbations. *Russian Math. Surveys*, 18(5):9–36, 1963.

[Arn63b] V. I. Arnol'd. Small denominators and problems of stability of motion in classical and celestial mechanics. *Russ. Math. Surveys*, 18:85–192, 1963.

[Arn64]  V. Arnold. Instability of dynamical systems with several degrees of freedom. *Sov. Math. Doklady*, 5:581–585, 1964.

[BB02]   M. Berti and P. Bolle. A functional analysis approach to Arnold diffusion. *Ann. Inst. H. Poincaré Anal. Non Linéaire*, 19(4):395–450, 2002.

[BBB02]  M. Berti, L. Biasco and P. Bolle. Optimal stability and instability results for a class of nearly integrable Hamiltonian systems. *Atti Accad. Naz. Lincei Cl. Sci. Fis. Mat. Natur. Rend. Lincei (9) Mat. Appl.*, 13(2):77–84, 2002.

[BBB03]  M. Berti, L. Biasco and P. Bolle. Drift in phase space: a new variational mechanism with optimal diffusion time. *J. Math. Pures Appl. (9)*, 82(6):613–664, 2003.

[BCV01]  U. Bessi, L. Chierchia and E. Valdinoci. Upper bounds on Arnold diffusion times via Mather theory. *J. Math. Pures Appl. (9)*, 80(1):105–129, 2001.

[Ber04]  P. Bernard. The dynamics of pseudographs in convex Hamiltonian systems, 2004. Preprint 04-313, `http://www.ma.utexas.edu/mp_arc/`.

[Bes96]  U. Bessi. An approach to Arnol'd's diffusion through the calculus of variations. *Nonlinear Anal.*, 26(6):1115–1135, 1996.

[BF04]   I. Baldomá and E. Fontich. Exponentially small splitting of invariant manifolds of parabolic points. *Mem. Amer. Math. Soc.*, 167(792):x–83, 2004.

[BK04]   J. Bourgain and V. Kaloshin. Diffusion for Hamiltonian perturbations of integrable systems in high dimensions. *J. Funct. Anal.*, to appear in 2004.

[BT99]   S. Bolotin and D. Treschev. Unbounded growth of energy in nonautonomous Hamiltonian systems. *Nonlinearity*, 12(2):365–388, 1999.

[Car81]  J. R. Cary. Lie transform perturbation theory for Hamiltonian systems. *Phys. Rep.*, 79(2):129–159, 1981.

[CFL03]  X. Cabré, E. Fontich and R. de la Llave. The parameterization method for invariant manifolds. I, II. *Indiana Univ. Math. J.*, 52(2):283–328, 329–360, 2003.

[CG94]   L. Chierchia and G. Gallavotti. Drift and diffusion in phase space. *Ann. Inst. H. Poincaré Phys. Théor.*, 60(1):144, 1994.

[CG98]   L. Chierchia and G. Gallavotti. Erratum drift and diffusion in phase space. *Ann. Inst. H. Poincaré Phys. Théor.*, 68:135, 1998.

[CG03]   J. Cresson and C. Guillet. Periodic orbits and Arnold diffusion. *Discrete Contin. Dyn. Syst.*, 9(2):451–470, 2003.

[Chi79]  B. Chirikov. A universal instability of many-dimensional oscillator systems. *Phys. Rep.*, 52(5):264–379, 1979.

[CLSV85] B. V. Chirikov, M. A. Lieberman, D. L. Shepelyansky and F. M. Vivaldi. A theory of modulational diffusion. *Phys. D*, 14(3):289–304, 1985.

[CP02]   G. Contreras and G. P. Paternain. Connecting orbits between static classes for generic Lagrangian systems. *Topology*, 41(4):645–666, 2002.

[CSUZ89] A. A. Chernikov, R. Z. Sagdeev, D. A. Usikov and G. M. Zaslavsky. *Weak chaos and structures*, volume 8 of *Soviet Scientific Reviews, Section C: Mathematical Physics Reviews*. Harwood Academic Publishers, Chur, 1989. ISBN 3-7186-4865-2.

[CV89]   B. V. Chirikov and V. V. Vecheslavov. How fast is the Arnold's diffusion? Technical Report 98-72, Inst. Plasma Phys., Novosibirsk, 1989.

[CY04a]  C.-Q. Cheng and J. Yan. Arnold diffusion in Hamiltonian systems: 1 a priori unstable case, 2004. Preprint 04-265, http://www.ma.utexas.edu/mp_arc/.

[CY04b]  C.-Q. Cheng and J. Yan. Existence of diffusion orbits in a priori unstable Hamiltonian systems, 2004.

[DG96]   A. Delshams and P. Gutiérrez. Effective stability and KAM theory. *J. Differential Equations*, 128(2):415–490, 1996.

[DG00]   A. Delshams and P. Gutiérrez. Splitting potential and the Poincaré-Melnikov method for whiskered tori in Hamiltonian systems. *J. Nonlinear Sci.*, 10(4):433–476, 2000.

[DG01]   A. Delshams and P. Gutiérrez. Homoclinic orbits to invariant tori in Hamiltonian systems. In C. K. R. T. Jones and A. I. Khibnik, editors, *Multiple-time-scale dynamical systems (Minneapolis, MN, 1997)*, pages 1–27. Springer, New York, 2001.

[DLC83]  R. Douady and P. Le Calvez. Exemple de point fixe elliptique non topologiquement stable en dimension 4. *C. R. Acad. Sci. Paris Sér. I Math.*, 296(21):895–898, 1983.

[DLS00]  A. Delshams, R. de la Llave and T. Seara. A geometric approach to the existence of orbits with unbounded energy in generic periodic perturbations by a potential of generic geodesic flows of $\mathbb{T}^2$. *Comm. Math. Phys.*, 209(2):353–392, 2000.

[DLS03]  A. Delshams, R. de la Llave and T. M. Seara. A geometric mechanism for diffusion in Hamiltonian systems overcoming the large gap problem: announcement of results. *Electron. Res. Announc. Amer. Math. Soc.*, 9:125–134, 2003.

[DLS04]  A. Delshams, R. de la Llave and T. M. Seara. Geometric properties of the scattering map to a normally hyperbolic manifold, 2004. Preprint.

[DLS05]  A. Delshams, R. de la Llave and T. Seara. Orbits of unbounded energy in generic quasiperiodic perturbations of geodesic flows of certain manifolds, To appear, 2005. MP_ARC 04-280.

[Dou88]  R. Douady. Stabilité ou instabilité des points fixes elliptiques. *Ann. Sci. École Norm. Sup. (4)*, 21(1):1–46, 1988.

[Eli94]  L. H. Eliasson. Biasymptotic solutions of perturbed integrable Hamiltonian systems. *Bol. Soc. Brasil. Mat. (N.S.)*, 25(1):57–76, 1994.

[EMR01]  R. W. Easton, J. D. Meiss and G. Roberts. Drift by coupling to an anti-integrable limit. *Phys. D*, 156(3-4):201–218, 2001.

[Fen72]  N. Fenichel. Persistence and smoothness of invariant manifolds for flows. *Indiana Univ. Math. J.*, 21:193–226, 1971/1972.

[Fen77]  N. Fenichel. Asymptotic stability with rate conditions. II. *Indiana Univ. Math. J.*, 26(1):81–93, 1977.

[Fen74]  N. Fenichel. Asymptotic stability with rate conditions. *Indiana Univ. Math. J.*, 23:1109–1137, 1973/74.

[FL92]   C. Falcolini and R. de la Llave. A rigorous partial justification of Greene's criterion. *J. Statist. Phys.*, 67(3-4):609–643, 1992.

[FM00]   E. Fontich and P. Martín. Differentiable invariant manifolds for partially hyperbolic tori and a lambda lemma. *Nonlinearity*, 13(5):1561–1593, 2000.

[FM01]   E. Fontich and P. Martín. Arnold diffusion in perturbations of analytic integrable Hamiltonian systems. *Discrete Contin. Dynam. Systems*, 7(1):61–84, 2001.

[FM03]   E. Fontich and P. Martín. Hamiltonian systems with orbits covering densely submanifolds of small codimension. *Nonlinear Anal.*, 52(1):315–327, 2003.

[FS90a]  E. Fontich and C. Simó. Invariant manifolds for near identity differentiable maps and splitting of separatrices. *Ergodic Theory Dynam. Systems*, 10(2):319–346, 1990.

[FS90b]  E. Fontich and C. Simó. The splitting of separatrices for analytic diffeomorphisms. *Ergodic Theory Dynam. Systems*, 10(2):295–318, 1990.

[Gal94]  G. Gallavotti. Twistless KAM tori, quasi flat homoclinic intersections, and other cancellations in the perturbation series of certain completely integrable Hamiltonian systems. A review. *Rev. Math. Phys.*, 6(3):343–411, 1994.

[Gal99]  G. Gallavotti. Arnold's diffusion in isochronous systems. *Math. Phys. Anal. Geom.*, 1(4):295–312, 1998/99.

[GL05]   M. Gidea and R. de la Llave. Topological methods in the instability problem of hamiltonian systems. *Discrete Contin. Dynam. Systems*, To appear, 2005. MP_ARC #05-93.

[Gra74]  S. M. Graff. On the conservation of hyperbolic invariant tori for Hamiltonian systems. *J. Differential Equations*, 15:1–69, 1974.

[Hal97]  G. Haller. Universal homoclinic bifurcations and chaos near double resonances. *J. Statist. Phys.*, 86(5-6):1011–1051, 1997.

[Hal99]  G. Haller. *Chaos near resonance.* Springer-Verlag, New York, 1999. ISBN 0-387-98697-9.

[Her79]  M.-R. Herman. Sur la conjugaison différentiable des difféomorphismes du cercle à des rotations. *Inst. Hautes Études Sci. Publ. Math.*, (49):5–233, 1979.

[Her83]  M.-R. Herman. *Sur les courbes invariantes par les difféomorphismes de l'anneau. Vol. 1*, volume 103 of *Astérisque*. Société Mathématique de France, Paris, 1983.

[HL00]   A. Haro and R. de la Llave. New mechanisms for lack of equipartion of energy. *Phys. Rev. Lett.*, 89(7):1859–1862, 2000.

[HM82]   P. Holmes and J. Marsden. Melnikov's method and Arnol'd diffusion for perturbations of integrable Hamiltonian systems. *J. Math. Phys.*, 23(4):669–675, 1982.

[HPS77]  M. Hirsch, C. Pugh and M. Shub. *Invariant manifolds*, volume 583 of *Lecture Notes in Math.* Springer-Verlag, Berlin, 1977.

[JVMU99] G. H. F. D. J. Von Milczewski and T. Uzer. The Arnold web in atomic physics. In C. Simó, editor, *Hamiltonian Systems with Three or More Degrees of Freedom (S'Agaró, 1995)*, pages 499–503. Kluwer Acad. Publ., Dordrecht, 1999.

[Kal03]  V. Kaloshin. Geometric proofs of Mather's connecting and accelerating theorems. In *Topics in dynamics and ergodic theory*, volume 310 of *London Math. Soc. Lecture Note Ser.*, pages 81–106. Cambridge Univ. Press, Cambridge, 2003.

[KMV04]  V. Kaloshin, J. Mather and E. Valdinoci. Instability of totally elliptic points of symplectic maps in dimension 4. *Astérisque*, to appear in 2004.

[Las93]  J. Laskar. Frequency analysis for multi-dimensional systems. Global dynamics and diffusion. *Phys. D*, 67(1-3):257–281, 1993.

[Lla00]  R. de la Llave. Persistence of hyperbolic manifolds, 2000.

[Lla01]  R. de la Llave. A tutorial on KAM theory. In A. Katok, R. de la Llave and Y. Pesin, editors, *Smooth ergodic theory and its applications (Seattle, WA, 1999)*, pages 175–292. Amer. Math. Soc., Providence, RI, 2001.

[Lla02]  R. de la Llave. Orbits of unbounded energy in perturbations of geodesic flows by periodic potentials. A simple construction, 2002. Preprint.

[LM88]   P. Lochak and C. Meunier. *Multiphase Averaging for Classical Systems*, volume 72 of *Appl. Math. Sci.* Springer, New York, 1988.

[LMM86]  R. de la Llave, J. M. Marco and R. Moriyón. Canonical perturbation theory of
         Anosov systems and regularity results for the Livšic cohomology equation. *Ann.
         of Math. (2)*, 123(3):537–611, 1986.

[LO99]   R. de la Llave and R. Obaya. Regularity of the composition operator in spaces
         of Hölder functions. *Discrete Contin. Dynam. Systems*, 5(1):157–184, 1999.

[LR02]   A. Litvak-Hinenzon and V. Rom-Kedar. Resonant tori and instabilities in Hamil-
         tonian systems. *Nonlinearity*, 15(4):1149–1177, 2002.

[LT83]   M. A. Lieberman and J. L. Tennyson. Chaotic motion along resonance layers in
         near-integrable Hamiltonian systems with three or more degrees of freedom. In
         C. W. Horton, Jr. and L. E. Reichl, editors, *Long-time prediction in dynamics
         (Lakeway, Tex., 1981)*, pages 179–211. Wiley, New York, 1983.

[LW95]   R. de la Llave and C. E. Wayne. On Irwin's proof of the pseudostable manifold
         theorem. *Math. Z.*, 219(2):301–321, 1995.

[LW04]   R. de la Llave and C. Wayne. Whiskered and lower dimensional tori in nearly
         integrable Hamiltonian systems. *Math. Phys. Electron. J.*, 10:Paper 5, 45 pp.
         (electronic), 2004.

[Mat93]  J. N. Mather. Variational construction of connecting orbits. *Ann. Inst. Fourier
         (Grenoble)*, 43(5):1349–1386, 1993.

[Mat95]  J. Mather. Graduate course at Princeton, 95–96, and Lectures at Penn State,
         Spring 96, Paris, Summer 96, Austin, Fall 96, 1995.

[Mat02]  J. N. Mather. Arnold diffusion I: Announcement of results. *Preprint*, 2002.

[Mei92]  J. D. Meiss. Symplectic maps, variational principles, and transport. *Rev. Modern
         Phys.*, 64(3):795–848, 1992.

[Mey91]  K. R. Meyer. Lie transform tutorial. II. In K. R. Meyer and D. S. Schmidt,
         editors, *Computer aided proofs in analysis (Cincinnati, OH, 1989)*, volume 28
         of *IMA Vol. Math. Appl.*, pages 190–210. Springer, New York, 1991.

[Moe96]  R. Moeckel. Transition tori in the five-body problem. *J. Differential Equations*,
         129(2):290–314, 1996.

[Moe02]  R. Moeckel. Generic drift on Cantor sets of annuli. In A. Chenciner, R. Cushman
         and C. Robinson, editors, *Celestial mechanics (Evanston, IL, 1999)*, volume 292
         of *Contemp. Math.*, pages 163–171. Amer. Math. Soc., Providence, RI, 2002.

[MS04]   J.-P. Marco and D. Sauzin. Wandering domains and random walks in Gevrey
         near-integrable systems. *Ergodic Theory Dynam. Systems*, 24(5):1619–1666,
         2004.

[Neĭ81]  A. I. Neĭshtadt. Estimates in the Kolmogorov theorem on conservation of con-
         ditionally periodic motions. *J. Appl. Math. Mech.*, 45(6):1016–1025, 1981.

[Neĭ84]  A. I. Neĭshtadt. The separation of motions in systems with rapidly rotating
         phase. *J. Appl. Math. Mech.*, 48(2):133–139, 1984.

[Nie00]  L. Niederman. Dynamics around simple resonant tori in nearly integrable Hamil-
         tonian systems. *J. Differential Equations*, 161(1):1–41, 2000.

[Poi99]  H. Poincaré. *Les méthodes nouvelles de la mécanique céleste*, volume 1, 2, 3.
         Gauthier-Villars, Paris, 1892–1899.

[Pös82]  J. Pöschel. Integrability of Hamiltonian systems on Cantor sets. *Comm. Pure
         Appl. Math.*, 35(5):653–696, 1982.

[RS02]   P. H. Rabinowitz and E. W. Stredulinsky. A variational shadowing method.
         In A. Chenciner, R. Cushman and C. Robinson, editors, *Celestial mechanics
         (Evanston, IL, 1999)*, volume 292 of *Contemp. Math.*, pages 185–197. Amer.
         Math. Soc., Providence, RI, 2002.

[Sal04]  D. Salamon. The Kolmogorov-Arnold-Moser theorem. *Math. Phys. Electron. J.*,
         10:Paper 3, 37 pp. (electronic), 2004.

[Sor02]  A. Sorrentino. *Sulle soluzioni quasi-periodiche di sistemi Hamiltoniani differen-
         ziabili*. Ph.D. thesis, Univ. di Roma Tre, 2002.

[Sva80]    N. V. Svanidze. Small perturbations of an integrable dynamical system with an integral invariant. *Trudy Mat. Inst. Steklov.*, 147:124–146, 204, 1980. English translation: *Proc. Steklov Inst. Math.*, 1981, no. 2.

[SZF95]    M. F. Shlesinger, G. M. Zaslavsky and U. Frisch, editors. *Lévy flights and related topics in physics.* Springer-Verlag, Berlin, 1995. ISBN 3-540-59222-9.

[Ten82]    J. Tennyson. Resonance transport in near-integrable systems with many degrees of freedom. *Phys. D*, 5(1):123–135, 1982.

[TLL80]    J. L. Tennyson, M. A. Lieberman and A. J. Lichtenberg. Diffusion in near-integrable Hamiltonian systems with three degrees of freedom. In M. Month and J. C. Herrera, editors, *Nonlinear dynamics and the beam-beam interaction (Sympos., Brookhaven Nat. Lab., New York, 1979)*, pages 272–301. Amer. Inst. Physics, New York, 1980.

[Tre91]    D. V. Treshchëv. A mechanism for the destruction of resonance tori in Hamiltonian systems. *Math. USSR-Sb.*, 68(1):181–203, 1991.

[Tre94]    D. V. Treshev. Hyperbolic tori and asymptotic surfaces in Hamiltonian systems. *Russian J. Math. Phys.*, 2(1):93–110, 1994.

[Tre02a]   D. Treschev. Multidimensional symplectic separatrix maps. *J. Nonlinear Sci.*, 12(1):27–58, 2002.

[Tre02b]   D. Treschev. Trajectories in a neighbourhood of asymptotic surfaces of a priori unstable Hamiltonian systems. *Nonlinearity*, 15(6):2033–2052, 2002.

[Tre04]    D. Treschev. Evolution of slow variables in a priori unstable hamiltonian systems. *Nonlinearity*, 17(5):1803–1841, 2004.

[Val00]    E. Valdinoci. Families of whiskered tori for a-priori stable/unstable Hamiltonian systems and construction of unstable orbits. *Math. Phys. Electron. J.*, 6:Paper 2, 31 pp. (electronic), 2000.

[Xia98]    Z. Xia. Arnold diffusion: a variational construction. In *Proceedings of the International Congress of Mathematicians, Vol. II (Berlin, 1998)*, Extra Vol. II, pages 867–877 (electronic). 1998.

[Zas02]    G. M. Zaslavsky. Chaos, fractional kinetics, and anomalous transport. *Phys. Rep.*, 371(6):461–580, 2002.

[Zeh76]    E. Zehnder. Generalized implicit function theorems with applications to some small divisor problems/II. *Comm. Pure Appl. Math.*, 29:49–111, 1976.

# Editorial Information

To be published in the *Memoirs*, a paper must be correct, new, nontrivial, and significant. Further, it must be well written and of interest to a substantial number of mathematicians. Piecemeal results, such as an inconclusive step toward an unproved major theorem or a minor variation on a known result, are in general not acceptable for publication. Papers appearing in *Memoirs* are generally at least 80 and not more than 200 published pages in length. Papers less than 80 or more than 200 published pages require the approval of the Managing Editor of the Transactions/Memoirs Editorial Board.

As of September 30, 2005, the backlog for this journal was approximately 14 volumes. This estimate is the result of dividing the number of manuscripts for this journal in the Providence office that have not yet gone to the printer on the above date by the average number of monographs per volume over the previous twelve months, reduced by the number of volumes published in four months (the time necessary for preparing a volume for the printer). (There are 6 volumes per year, each containing at least 4 numbers.)

A Consent to Publish and Copyright Agreement is required before a paper will be published in the *Memoirs*. After a paper is accepted for publication, the Providence office will send a Consent to Publish and Copyright Agreement to all authors of the paper. By submitting a paper to the *Memoirs*, authors certify that the results have not been submitted to nor are they under consideration for publication by another journal, conference proceedings, or similar publication.

## Information for Authors

*Memoirs* are printed from camera copy fully prepared by the author. This means that the finished book will look exactly like the copy submitted.

The paper must contain a *descriptive title* and an *abstract* that summarizes the article in language suitable for workers in the general field (algebra, analysis, etc.). The *descriptive title* should be short, but informative; useless or vague phrases such as "some remarks about" or "concerning" should be avoided. The *abstract* should be at least one complete sentence, and at most 300 words. Included with the footnotes to the paper should be the 2000 *Mathematics Subject Classification* representing the primary and secondary subjects of the article. The classifications are accessible from `www.ams.org/msc/`. The list of classifications is also available in print starting with the 1999 annual index of *Mathematical Reviews*. The Mathematics Subject Classification footnote may be followed by a list of *key words and phrases* describing the subject matter of the article and taken from it. Journal abbreviations used in bibliographies are listed in the latest *Mathematical Reviews* annual index. The series abbreviations are also accessible from `www.ams.org/publications/`. To help in preparing and verifying references, the AMS offers MR Lookup, a Reference Tool for Linking, at `www.ams.org/mrlookup/`. When the manuscript is submitted, authors should supply the editor with electronic addresses if available. These will be printed after the postal address at the end of the article.

**Electronically prepared manuscripts.** The AMS encourages electronically prepared manuscripts, with a strong preference for $\mathcal{A}\mathcal{M}\mathcal{S}$-LaTeX. To this end, the Society has prepared $\mathcal{A}\mathcal{M}\mathcal{S}$-LaTeX author packages for each AMS publication. Author packages include instructions for preparing electronic manuscripts, the *AMS Author Handbook*, samples, and a style file that generates the particular design specifications of that publication series. Though $\mathcal{A}\mathcal{M}\mathcal{S}$-LaTeX is the highly preferred format of TeX, author packages are also available in $\mathcal{A}\mathcal{M}\mathcal{S}$-TeX.

Authors may retrieve an author package from e-MATH starting from `www.ams.org/tex/` or via FTP to `ftp.ams.org` (login as `anonymous`, enter username as password, and type `cd pub/author-info`). The *AMS Author Handbook* and the *Instruction Manual* are available in PDF format following the author packages link from `www.ams.org/tex/`. The author package can be obtained free of charge by sending email to `pub@ams.org` (Internet) or from the Publication Division, American Mathematical Society, 201 Charles St., Providence, RI 02904, USA. When requesting an author package, please specify $\mathcal{A}\mathcal{M}\mathcal{S}$-LaTeX or $\mathcal{A}\mathcal{M}\mathcal{S}$-TeX, Macintosh or IBM (3.5) format, and the publication in which your paper will appear. Please be sure to include your complete mailing address.

**Sending electronic files.** After acceptance, the source file(s) should be sent to the Providence office (this includes any TeX source file, any graphics files, and the DVI or PostScript file).

Before sending the source file, be sure you have proofread your paper carefully. The files you send must be the EXACT files used to generate the proof copy that was accepted for publication. For all publications, authors are required to send a printed copy of their paper, which exactly matches the copy approved for publication, along with any graphics that will appear in the paper.

TeX files may be submitted by email, FTP, or on diskette. The DVI file(s) and PostScript files should be submitted only by FTP or on diskette unless they are encoded properly to submit through email. (DVI files are binary and PostScript files tend to be very large.)

Electronically prepared manuscripts can be sent via email to `pub-submit@ams.org` (Internet). The subject line of the message should include the publication code to identify it as a Memoir. TeX source files, DVI files, and PostScript files can be transferred over the Internet by FTP to the Internet node `e-math.ams.org` (130.44.1.100).

**Electronic graphics.** Comprehensive instructions on preparing graphics are available at `www.ams.org/jourhtml/graphics.html`. A few of the major requirements are given here.

Submit files for graphics as EPS (Encapsulated PostScript) files. This includes graphics originated via a graphics application as well as scanned photographs or other computer-generated images. If this is not possible, TIFF files are acceptable as long as they can be opened in Adobe Photoshop or Illustrator. No matter what method was used to produce the graphic, it is necessary to provide a paper copy to the AMS.

Authors using graphics packages for the creation of electronic art should also avoid the use of any lines thinner than 0.5 points in width. Many graphics packages allow the user to specify a "hairline" for a very thin line. Hairlines often look acceptable when proofed on a typical laser printer. However, when produced on a high-resolution laser imagesetter, hairlines become nearly invisible and will be lost entirely in the final printing process.

Screens should be set to values between 15% and 85%. Screens which fall outside of this range are too light or too dark to print correctly. Variations of screens within a graphic should be no less than 10%.

**Inquiries.** Any inquiries concerning a paper that has been accepted for publication should be sent directly to the Electronic Prepress Department, American Mathematical Society, 201 Charles St., Providence, RI 02904, USA.

# Titles in This Series

For a complete list of titles in this series, visit the
AMS Bookstore at **www.ams.org/bookstore/**.